Asphalt and Asphalt Mixtures

Edited by Haitao Zhang

Published in London, United Kingdom

IntechOpen

Supporting open minds since 2005

Asphalt and Asphalt Mixtures
http://dx.doi.org/10.5772/intechopen.80164
Edited by Haitao Zhang

Contributors
Yıldırım İsmail İsmail Tosun, Paula Cristina Arroyo-Martínez, Rey Omar Adame Hernández, Israel Sandoval-Navarro, Ignacio Roberto Cremades-Ibáñez, Norma Angélica Sánchez-Flores, Mohammadsoroush Tafazzoli, Shahab Hasaninasab, Abayomi Modupe, Wilmar Darío Fernández-Gómez, Carlos Alfonso Cuadro Causil, Jorge Iván Osorio Esquivel, Fredy Alberto Reyes Lizcano, Haitao Zhang

First published in London, United Kingdom, 2019 by IntechOpen
IntechOpen is the global imprint of INTECHOPEN LIMITED, registered in England and Wales, registration number: 11086078, 7th floor, 10 Lower Thames Street, London, EC3R 6AF, United Kingdom
Printed in Croatia

British Library Cataloguing-in-Publication Data
A catalogue record for this book is available from the British Library

Additional hard and PDF copies can be obtained from orders@intechopen.com

Asphalt and Asphalt Mixtures
Edited by Haitao Zhang
p. cm.
Print ISBN 978-1-78984-768-0
Online ISBN 978-1-78984-769-7
eBook (PDF) ISBN 978-1-83968-109-7

For EU product safety concerns:
IN TECH d.o.o., Prolaz Marije Krucifikse Kozulić 3, 51000 Rijeka, Croatia, info@intechopen.com or visit our website at intechopen.com.

Meet the editor

Professor Haitao Zhang graduated from Tongji University, Shanghai in China with a major in civil engineering. He currently works at the Northeast Forestry University as a professor. His research interests include pavement design theory, asphalt mix, and pavement management and maintenance.

Contents

Preface XI

Section 1
Introduction 1

Chapter 1 3
Introductory Chapter: Asphalt and Asphalt Mixture
by Haitao Zhang

Section 2
Functional Asphalt Mixture 15

Chapter 2 17
Warm Mix Asphalt
by Paula Cristina Arroyo-Martínez, Rey Omar Adame-Hernández, Ignacio Roberto Cremades-Ibáñez, Israel Sandoval-Navarro and Norma Angélica Sánchez-Flores

Chapter 3 27
The Effects of Using Nano-Silica in Cold-In Place Asphalt with Emulsified Bitumen
by Shahab Hasaninasab

Chapter 4 35
Incorporating Sustainable Practices in Asphalt Industry
by Mohammadsoroush Tafazzoli

Section 3
Asphalt Aging and Rejuvenation 55

Chapter 5 57
The Evaluation of Asphalt Mixture Mastic as an Aging Indicator
by Carlos Alfonso Cuadro Causil, Wilmar Darío Fernández-Gómez, Jorge Iván Osorio Esquivel and Fredy Alberto Reyes Lizcano

Section 4
Asphalt Pavement 77

Chapter 6 79
Microwave Processing of Mixtures of Asphalt and Şırnak Asphaltite
Slime - Development of The Compression Ability, Shear and Tension Strength
by Yıldırım İsmail Tosun

Chapter 7 105
Polymer-Modified Bio-Asphalt: A Sustainable Panacea to Greenhouse
Gas Emissions
by Modupe Abayomi Emmanuel

Preface

Road pavement types include asphalt pavement, cement concrete pavement, and earth pavement. With the rapid growth of economy, people continue to improve and innovate in the construction of roads. In recent years, asphalt has been used to build most expressways or first-grade highways. People are also constantly renovating a large number of old asphalt pavements to improve the pavement quality and make it flatter, so as to extend the service life of the pavement. At present, the main problems in the research of asphalt and asphalt mixture are how to further improve the durability of asphalt pavement and the prevention and control of asphalt pavement diseases. These problems are the key problems that affect the service life of asphalt pavement. The proposal and research of these problems will further promote the application of asphalt pavement. The traditional asphalt pavement mainly emphasizes the mechanical properties of the pavement. With the continuous development of human society, asphalt pavement should not only meet the requirements of mechanical properties, but also meet the functional requirements, such as pavement drainage, noise reduction, and so on. Asphalt pavement, with both a mechanical performance index and functional index, has gradually become a research hotspot.

Haitao Zhang
Professor,
College of Civil Engineering,
Northeast Forestry University,
Harbin, PR China

Section 1

Introduction

Chapter 1

Introductory Chapter: Asphalt and Asphalt Mixture

Haitao Zhang

1. Introduction

As the most important material for pavement construction, asphalt has always been valued. In 1987, the US Congress established the Strategic Highway Research Program (SHRP) to improve the performance and durability of US highways. The PG performance grading proposed by the US SHRP program is being learned and used by many countries in the world as a new technology. As a kind of polymer material, the microstructure of asphalt is very complicated. Therefore, the microstructure is not used as the evaluation index and standard, but the characteristic index and standard with theological basis are used in the classification of asphalt. The asphalt standards adopted by various countries also have their own characteristics. Some asphalt indexes are many, such as SHRP asphalt performance specifications and the European CEN asphalt new standards, and some have only a few indicators, such as the new Canadian asphalt standard and the Australian asphalt standard. The research on technical indexes and standards of asphalt in China has been carried out for many years, mainly referring to the indexes and standards of foreign countries. At the same time, combined with the test and material characteristics of China, some modifications have been made. The asphalt mixture is a multiphase dispersion system with space network structure, which is composed of aggregate and binder, and the mechanical strength of the asphalt mixture is mainly composed of the internal friction resistance and the embedding force between the mineral particles and the adhesive force between the asphalt cement and the mineral material. According to the proportion of the embedded structure and the dense structure in the asphalt mixture, the asphalt mixture structure is generally divided into three types: a suspended dense structure, a skeleton void structure, and a skeleton dense structure. Although the performance of a single material in asphalt mixture plays a very important role in the performance of asphalt mixture, the combination characteristics of asphalt and aggregate composition system in asphalt mixture have greater influence on the performance of asphalt mixture. The properties of asphalt mixture include permanent deformation, fatigue cracking, and low temperature cracking [1, 2].

There are three methods for asphalt mix design, namely Marshall design method, Hveem design method, and Superpave design method. The asphalt binder used in the Marshall design method of China asphalt mixture is based on the asphalt technical index system of JTG F40-2011. The system evaluates the performance of road asphalt with three major indexes of asphalt. According to different design traffic volume and different natural factors (temperature, etc.), asphalt binder is selected. A new asphalt mixture design system, superior performance asphalt pavement (Superpave), has been developed by the Strategic Highway Research Program in the United States. Compared with Marshall design method, Superpave design method is a completely different asphalt mixture composition design system, which includes new test equipment, material selection and design, test standards, and so on. With the continuous development of human society, people's functional requirements for pavements are increasing. How to further enhance the mechanical performance of asphalt pavements while taking into account functional indicators (drainage, noise reduction, anti-sliding, etc.) has become a problem to be solved.

Asphalt and asphalt mixtures mainly include (**Figure 1**):

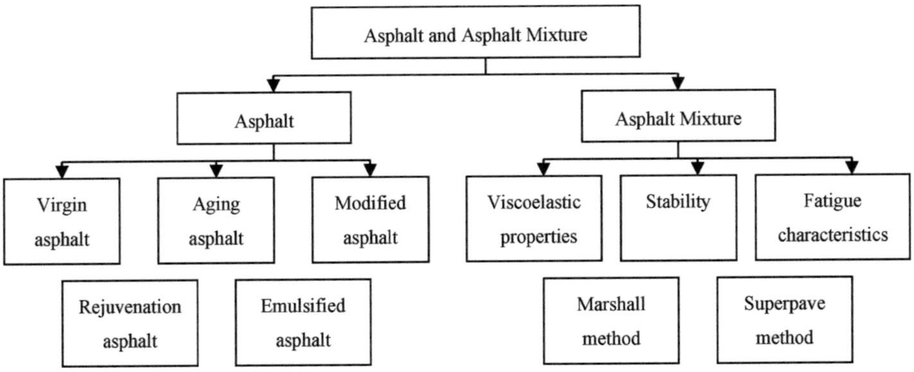

Figure 1.
Classification of asphalt and asphalt mixture.

2. Modified asphalt and mixture

As early as the early nineteenth century, rubber powder was first applied to asphalt. Rubber-modified asphalt first appeared in people's field of vision. In the middle of the nineteenth century, the vulcanization of asphalt was first proposed. It was found that the vulcanization of asphalt can have an important impact on the high temperature properties of asphalt. In the early twentieth century, France built the first road using rubber-modified asphalt. In the 1950s, the United States and Japan also conducted in-depth exploration of rubber-modified asphalt and verified it through the paving of many test roads. In the middle of the twentieth century, SBS-modified asphalt was studied, and there was a major breakthrough in 30 years. Nowadays SBS-modified asphalt has become popular; in the 1960s, styrene-butadiene (SBR)-modified asphalt was successfully developed. It was found that SBR-modified asphalt has a good improvement effect on low temperature stability. In the mid-1980s, China began research on SBS asphalt, from which SBS-modified asphalt appeared on the Chinese highway as the most extensive asphalt binder. At this stage, China's SBS-modified asphalt production has exceeded 500,000 tons; in the 1990s, rubber powder-modified asphalt began to appear in China, and in 1993, the first rubber-modified asphalt pavement was paved in Shenyang. Since then, modified asphalt has appeared in China's high-grade highways. During the 2008 Olympic Games, China launched the theme of Green Olympics, High-tech Olympics, and Humanistic Olympics. China put forward new ideas on the disposal of used rubber powder and organized a series of scientific research related to waste rubber powder [3].

The so-called modified asphalt (mixture) refers to an asphalt binder made of rubber, resin, high molecular polymer, natural asphalt, ground rubber powder, or other external admixture (modifier), thereby improving the asphalt or the performance of the asphalt mixture. A modifier refers to a natural or artificial organic or inorganic material added to asphalt (mixture). At the same time, with the development of the times, a single modifier has a certain effect on improving the durability of the asphalt mixture, but its ability to resist aging, plastic deformation, fatigue, and other durability is relatively weak, so the modification effect of a single modifier on asphalt cannot meet the requirements of existing pavement performance. Therefore, composite-modified asphalt has emerged. Composite-modified asphalt mainly refers to multiple modification of matrix asphalt.

Regarding the classification of modified asphalt, there is no uniform classification standard in the world, and it is currently classified mainly according to the variety of modifiers used. The modified asphalt can be roughly divided into three categories according to the different modifiers:

1. Rubber and thermoplastic elastomer-modified asphalt including natural rubber-modified asphalt, SBS-modified asphalt, styrene-butadiene rubber-modified asphalt, neoprene-modified asphalt, butadiene rubber-modified asphalt, butyl rubber-modified asphalt, waster rubber- and recycled rubber-modified asphalt, other rubber-modified asphalt (such as ethylene propylene rubber, nitrile rubber, etc.).

2. Plastic- and synthetic resin class-modified asphalt including polyethylene-modified asphalt, ethylene-vinyl acetate polymer-modified asphalt, polystyrene-modified asphalt, coumarin resin-modified asphalt, epoxy resin-modified asphalt, α-olefin random polymer-modified asphalt.

3. Resonance-type polymer-modified asphalt: the asphalt is modified by adding two or more polymers to the asphalt at the same time. The two or more kinds of polymers referred to herein may be two separate high molecular polymers or may be a so-called polymer alloy which has been previously blended to form a polymer interpenetrating network.

As a complex polymer hydrocarbon, asphalt exhibits typical elastic-visco-plasticity under certain temperature and load. The main purpose of adding modifiers is to improve the high and low temperature properties of asphalt mixtures (anti-rutting, anti-fatigue, anti-aging, resistance to low temperature cracking, etc.). The mechanism of action of the modifier can be summarized as follows: the modifier is sufficiently miscible with the asphalt. On the basis of this, the modifier adsorbs the light components in the asphalt and swells, the swelling modifier, and the rest of the asphalt. The components interact to form a new resulting system, combined with the inherent properties of the modifier itself to provide a corresponding improvement in asphalt performance. At the same time, in the grinding process and under the action of the stabilizer or the catalyst, the chain scission and the cross-linking reaction occur, and some network structures are formed, so that the viscosity and storage stability of the modified asphalt are improved. Therefore, the modified asphalt is a new structural material that has both the basic characteristics of asphalt and polymer.

3. Asphalt aging and rejuvenation

3.1 Mechanism of asphalt aging

Asphalt aging refers to a series of volatilization, oxidation, polymerization, and other changes under the action of environmental factors (heat, oxygen, sunlight, and water). In the aging process, the light components in the asphalt are volatilized and absorbed, the molecular structure is changed, the asphalt is hard and brittle, and the adhesive property is reduced to generate cracks. Asphalt aging is mainly manifested by increasing softening point and decreasing penetration. Although the aging of asphalt will enhance the rut resistance of asphalt pavement at high temperature, the low temperature performance and fatigue resistance of asphalt pavement will be greatly reduced; thus the adhesion and bonding ability of asphalt pavement will become worse. Asphalt aging is a gradual process, and its rate directly affects the service life of pavement, so it is the main factor affecting the durability of asphalt pavement [4].

3.2 Classification of asphalt aging

3.2.1 Short-term aging

1. Transport and storage process. The asphalt transportation is about 170°C, the number of the process asphalt is large and the depth is large, so that the contact air is small and the aging degree is small.

2. The aging of the mixing process. The aging of asphalt in this process will be further aggravated because the asphalt in this process is fully in contact with many factors, such as air.

3. The aging of the construction period. Asphalt from transportation to construction, temperature reduction and recovery to natural temperature, site paving, rolling, so that asphalt aging further development.

3.2.2 Long-term aging

The long-term aging of asphalt is a complex and slow process, continuous and uninterrupted, and the action factors are complex. The aging degree is further increased with the influence of vehicle load and temperature.

3.3 Aging of asphalt pavement and its influencing factors

3.3.1 External reasons

1. The service life of the road surface. The longer the service life of asphalt pavement, the worse the aging.

2. The asphalt pavement has different depths. The study shows that the depth of the asphalt pavement with severe aging is generally only in the range of 0.5–1 cm.

3. Asphalt pavement location is different. Aging is more severe in places where there are more axles than axle rolling less.

4. Grading type of mixture. Under the same porosity, the air permeability of intermittent graded mixture is smaller than that of continuous graded mixture.

3.3.2 The internal reasons

The aromatic components (Ar) in asphalt and oxygen in air are oxidized to form colloidal (R). In the process of aging, colloids can easily be transformed into asphaltene (A) components with poor relative molecular weight through polymerization and condensation.

3.4 Asphalt regeneration mechanism

3.4.1 Regeneration mechanism of old asphalt materials

There are two theories on the mechanism of old asphalt regeneration:

1. Compatibility theory: it is considered that the aging of asphalt is due to the decrease of the compatibility of each component in the asphalt colloid system, which leads to the increase of the solubility parameter difference of the components. It is considered that the solubility parameter difference can be reduced by adding a certain regenerated agent and the asphalt can recover to (or even exceed) the original property [5].

2. The theory of component regulation: it is considered that the pavement performance of asphalt decreases due to the migration of components and the uncoordinated proportion of each component after aging. It is considered that the asphalt can be restored to its original properties by adding regenerated agent to adjust its components. Therefore, in order to restore the original performance of the aged asphalt, it is necessary to compare the components of the aged asphalt with the original asphalt and add the missing components (i.e., adding asphalt recycling agent) to the aging asphalt so that the components can be re-coordinated.

3.4.2 Asphalt regeneration method

1. New and old asphalt mixing and regeneration. Mix the new asphalt with higher grade with the old asphalt, and mix the softer new asphalt with the aged old asphalt. The mixed asphalt meets the road asphalt standard.

2. Regeneration agent regeneration. Adding proper amount of recycled additive to the old asphalt can not only adjust the viscosity of the old asphalt but also supplement the lost chemical components of the old asphalt, restore the performance of the original asphalt, and even exceed the performance of the original asphalt.

3. Mixed regeneration. When the new asphalt is added and the recycled agent is mixed, the recycled material can obtain better performance.

4. Functional asphalt mixture/pavement

With the continuous development of social civilization and the continuous improvement of road engineering construction level, people's understanding and functional needs of road are increasing day by day. Since the twenty-first century, the emergence of new functional materials and the development of cross-discipline have provided a powerful support for the design and construction of all kinds of environment-friendly functional pavement and further broaden the research field of pavement materials. Constructing environment-friendly functional pavement, further improving road area environment, has become the frontier direction of road engineering material discipline development. As a part of road structure which acts directly with driving load and natural environment, the traditional requirements of traditional pavement are load bearing, leveling, safety, and durability.

In addition to the basic requirements to ensure normal road driving conditions, modern pavement should also have new environmental protection functions, such as water permeability, noise reduction, low heat absorption, car exhaust pollution, and so on, from the point of view of improving human settlements. Pervious asphalt concrete, generally used as a wear layer for pavement. Many large-scale tests of permeable asphalt concrete were carried out in 1967 to determine the applicability of permeable asphalt macadam to heavy traffic roads. The research report of Szatkowski and Brown pointed out that the asphalt macadam mixed with large particle-size aggregate generally has the maximum voidage permeability and texture depth, and the material mixed with 19 mm aggregate has a good effect on reducing the water fog caused by traffic. By reducing a large amount of water fog, the reflection phenomenon on the road surface can be eliminated, so that the road signs can be kept in high visibility, which is beneficial to traffic safety. At the same time, the rainwater falling on the surface of the permeable asphalt gravel can flow out of the pavement through the pores inside the surface layer, so that it also acts as a drainage layer. The rainwater can be allowed to permeate out of the surface layer instead of being accumulated on the surface of the road to form a water film or runoff, thereby avoiding the phenomenon of water drift generated when the vehicle is traveling at a high speed on a general road surface in rainy days. The surface of the road has a rough macroscopic texture, and the tire surface strengthens the contact with the road surface when the vehicle is driving at high speed in rainy days, which helps to maintain good anti-slip ability. The noise generated by vehicles on porous pavement is much lower than that of ordinary asphalt pavement with the same anti-skid degree, 3–4 dB (A), lower when dry, and 78 dB (A) lower in wet condition. Another advantage of permeable asphalt macadam pavement is to reduce the rolling resistance of vehicle tires and save a lot of fuel.

Although permeable asphalt pavements have many distinct advantages, there are some disadvantages that must be overcome, for example, how to balance their mechanical properties and functional specifications. The allowable variation range of the asphalt content of the permeable asphalt mixture is small, and if the asphalt content is too low, the road surface may be damaged in advance. On the other hand, if the asphalt content is too high, the asphalt will flow in the process of transportation or paving, resulting in uneven asphalt content. At present, the mixing ratio design method of the porous asphalt mixture is to adopt a leak test and a scattering test. The existence of voids makes the pavement materials more exposed to water, air, ultraviolet, and other environmental factors. The formation mechanism of contact interface strength between aggregates and binders is systematically studied, which is of great significance to improve the bonding strength and durability of aggregates and binders. The composition characteristics and functional objectives of porous pavement materials determine that the service environment is relatively complex, and it is necessary to meet the function and balance of multiple physical fields, such as sound absorption, water permeability, heat transfer, and so on. Therefore, it is very important to study its road performance under the action of multi-physical field coupling [6].

5. Microstructure behavior of asphalt

In the process of asphalt research in recent years, the macroscopic study of asphalt has always been a leading position. Therefore, the proportion of the study on the microstructure characteristics of asphalt will gradually increase in the future research. Moreover, the microstructure of asphalt is extremely complex, and the domestic and foreign researches on it are not perfect, so increasing the research

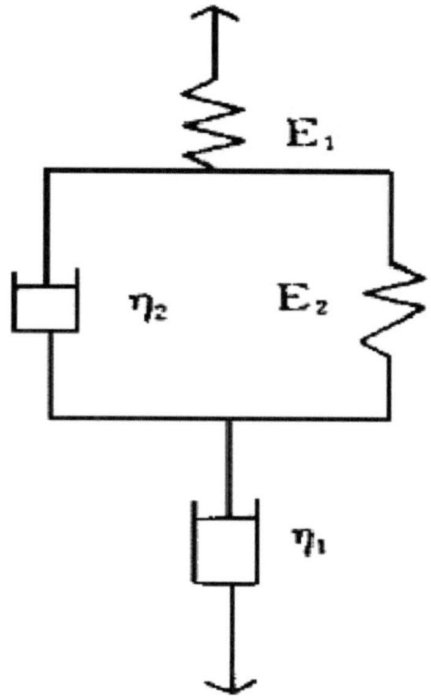

Figure 2.
Burgers model.

on the microstructure of asphalt has great significance for the development of asphalt. Rheological model, infrared spectrum, and atomic force microscope are used to study the microstructure of asphalt. Structural characteristics of asphalt are generally studied by using existing rheological models, including Maxwell model, Kelvin model, Burgers model, etc. However, Burgers model is the most common one that can demonstrate the viscoelasticity of asphalt in a more comprehensive way. And it is composed by a set of Maxwell models and a set of Kelvin models (shown in **Figure 2**). The constitutive equation is as follows:

$$x(t) = e\left[\frac{1}{E_1} + \frac{1}{Z_1}t - \frac{1}{Z_2}\left(2 - e^{\frac{-E_2}{Z_2}t}\right)\right]$$

where e is load, t is time, E_1 and E_2 are moduli of elasticity, and Z_1 and Z_2 are the viscosity coefficients.

According to the US SHRP program, when we understand two moduli of elasticity and two viscosity coefficients, we can understand the viscosity characteristics of different asphalt and thus understand the structural characteristics of asphalt. Infrared spectrum analysis can be used to analyze the microscopic mechanism of different modified asphalt and different aged asphalt. By scanning the changes of asphalt molecules by projection spectrum, the relationship between component content and characteristic absorption peak absorbance is analyzed, so as to determine the content of different components of different asphalt. Infrared spectrum analysis is common now because of its simple sample preparation and sampling, short test and analysis time, and high accuracy of measured results. The detection principle of AFM is that during the scanning process, the change in the height of the sample microsurface will change the micro-force (attraction or repulsion force) between the tip and the sample surface and the microcantilever will be deviated according to Hooke's law. Usually, this offset will cause the reflection number of the laser source irradiated on the back of the microcantilever to change, and it will be sensed by the photodiode. After processing, the surface topography of the sample, namely, the "peak structure," can be obtained. Through the observation of the peak structure of asphalt and image processing technology, the fractal dimension and other technical means are used to analyze the internal micro of asphalt. Different aging time and modifier will affect the area of "peak structure," which will also affect the microstructure characteristics of different asphalt.

In the development of modern asphalt, the research on modified asphalt has been more and more mature. More and more materials are used in the study of modified asphalt, from which a lot of valuable experience has been obtained. The asphalt modified by different materials has different effects. Among all modified asphalt, nanomaterial-modified asphalt is of great research value. In recent years, nanomaterials have been widely used in the field of road materials. Nano-modified asphalt is a kind of asphalt nanocomposite material, which mainly studies the performance of nanomaterials on road building materials. Many achievements have been made in the experimental research, practical application, and research methods of nanomaterial-modified asphalt at home and abroad, but there are still many deficiencies. Therefore, the preparation process of nanometer-modified asphalt should be improved, and new test methods and research methods should be explored to make the preparation process of nanometer-modified asphalt simple and the road performance good [7].

6. Asphalt pavement

Road pavement is divided into asphalt pavement, cement concrete pavement, and earth pavement. With the rapid growth of economy, people continue to improve and innovate in the construction of roads. In recent years, most of them use asphalt to build expressways or first-grade highways. People are also constantly renovating a large number of asphalt pavements built in the early stage to improve the pavement quality and make it more flat, so as to extend the service life of the pavement.

Asphalt pavement must have several basic conditions in the process of use:

1. It has some strength. The pavement will produce certain deformation under the action of load. If the allowable stress of pavement material is not enough to resist the stress generated by pavement structure layer, the pavement will produce failure cracking.

2. It has certain stability. Stability of asphalt pavement includes high and low temperature stability, water stability, etc. High temperature stability is shown in the ability to resist permanent deformation of asphalt pavement, and low temperature stability is shown in the ability to resist low temperature cracking of asphalt pavement. Water stability refers to the ability not to damage under the action of water and load.

3. Good durability. There are many factors that affect the durability of asphalt pavement, including temperature, humidity, material performance, load size, and so on, which will have an impact on the durability.

4. Good flatness. Good flatness can reduce the loss of the car but also can improve the speed of the car and the ride in the process of comfort and safety.

5. Good skid resistance. The car on the smooth road driving lack of wheel and road friction, easy to appear dangerous, especially in rainy or snowy days when suddenly start or brake, will produce idling or skid phenomenon, serious traffic accidents will occur. Therefore, it requires high skid resistance for the road surface.

With the continuous development and improvement of asphalt pavement, there are many kinds of modern asphalt pavement, such as functional asphalt pavement, smart pavement, and recycled pavement. Functional asphalt pavement refers to the pavement with different functions on the basis of the original materials, which gives different functions and functions to asphalt pavement, including high-modulus anti-rutting asphalt pavement, porous drainage noise reduction asphalt pavement, and flame-retardant asphalt pavement. The appearance of functional pavement improves the performance of asphalt pavement and makes a great contribution to the improvement of ecological environment and sustainable development in the future. From the development trend of asphalt pavement in China, these roads have a good development prospect.

Europe has given an open definition of the smart road, which means that the next generation of roads should be adaptive, automated, and resilient, with the ability to easily accommodate new technologies and adapt to climate change. Smart road is all kinds of information, technology and function of the integration of the body, have more access to the information and data processing method, technology and development of the service function will be constantly updated, all of which require a unified architecture to build efficient, flexible and economic system, intelligent pavement for designers to provide the necessary guidance and support. Therefore, the development of intelligent road surface will be an important branch of road surface development in the future.

Globally, asphalt mixture recycling has been used in practical engineering since the early twentieth century, but it was ignored until the oil crisis of 1973. The emergence of oil crisis makes asphalt pavement regeneration technology has a great development. By the second half of the twentieth century, asphalt pavement recycling technology has made a lot of research achievements. The United States, Britain, Germany, Japan, and other countries have prepared suitable for their own application of recycling technology manuals, guidelines and specifications. Compared with Europe and America, the research on asphalt pavement regeneration technology started relatively late in Asian countries except Japan. With the deepening understanding of the huge environmental and economic benefits brought by this technology, Asian countries accelerated the research on this technology, and now it has been widely used in all countries. Recycled asphalt mixture can be regarded as a special asphalt mixture, because it contains a certain proportion of old mixture; this part of the old aging material will have a certain impact on the performance of recycled mixture. This will make a certain contribution to the environmental protection cause in the future [8].

7. Conclusions

In recent years, with the rapid development of the world transportation industry, the proportion of asphalt pavement in road engineering is increasing. Therefore, while the demand for asphalt and asphalt mixture is increasing, the quality requirements for materials are also increasing. In particular, new materials and new technologies are constantly emerging, and the application technology, theory, and technical specifications of asphalt materials have made great progress and updates. Porous structural materials can achieve functions such as noise reduction, water permeability, temperature reduction, and decomposition of automobile exhaust, but the durability, functional improvement, and curing methods of various materials need further study. The change in understanding of the pavement from a single channel function to a comprehensive channel plus environmentally friendly function is not only due to the expansion of the pavement function, which will lead to an update of the pavement material and structural technology design concepts. The results will drive the progress of modern road engineering technology and the development of related social industries. At present, the main problems in the research of asphalt and asphalt mixture are how to further improve the durability of asphalt pavement and the prevention and control of asphalt pavement diseases. These problems are the key problems that affect the service life of asphalt pavement. The proposal and research of these problems will further promote the application of asphalt pavement. The traditional asphalt pavement mainly emphasizes the mechanical properties of the pavement. With the continuous development of human society, asphalt pavement should not only meet the requirements of mechanical properties but also meet the functional requirements, such as pavement drainage, noise reduction, and so on. Asphalt pavement with both mechanical performance index and functional index has gradually become a research hotspot.

Author details

Haitao Zhang
College of Civil Engineering, Northeast Forestry University, Harbin, China

*Address all correspondence to: zht6781@163.com

References

[1] Deng X. Subgrade and pavement engineering. Beijing, China: China Communication Press; 2000

[2] HWA. Superpave Mixture Design Guide. Westrack Forensic Consensus Report. FHWA-RD-01-052. USA. 2001

[3] Ministry of Transport of the People's Republic of China. Test specification for asphalt and bituminous mixtures for Highway Engineering (JTG E20-2011). Beijing: China Communications Press; 2011

[4] FHWA. Basic Asphalt Recycling Manual. Asphalt Recycling and Reclaiming Association. USA. 2001

[5] Austroads. Asphalt Recycling Guide. Sydney, Australia. 1997

[6] America Asphalt Institute. Superior Performing Asphalt Pavements (Superpave). Beijing, China: China Communication Press; 2005

[7] Kennedy TW. Superior Performing Asphalt Pavements (Superpave). The Product of the SHRP Asphalt Research Program. Strategic Highway Research Program. Washington DC, USA: SHRP-A-410 National Research Council; 1997

[8] Yu J, Jirong C, Benjing LI. Manual of Superior Performing Asphalt Pavements (Superpave). Beijing, China: China Communication Press; 2005

Section 2

Functional Asphalt Mixture

Chapter 2

Warm Mix Asphalt

Paula Cristina Arroyo-Martínez, Rey Omar Adame-Hernández, Ignacio Roberto Cremades-Ibáñez, Israel Sandoval-Navarro and Norma Angélica Sánchez-Flores

Abstract

Warm mix asphalt (WMA) is a technology that emerges to achieve environmental challenges of reducing greenhouse gas emissions. There are several technologies that allow diminishing the mixing and compaction temperature of the asphalt mixtures while improving workability. The benefits of using warm mix asphalt are not just environmental but also include better working conditions and the capability of introducing greater percentages of recycled materials into the mixture. Foamed asphalt is the most used technology to obtained warm mix asphalt in the United States of America (USA), and the performance of the resultant mixtures could be increased by controlling and improving the characteristics of the foam.

Keywords: foamed asphalt, warm mix asphalt, warm mix asphalt technologies, foamed warm mix asphalt additives, foam characteristics

1. Introduction

The implementation of engineering controls on asphalt pavement industries during 1997 became an important step in the environmentally friendly technologies in the industry. These controls included diminishing the greenhouse gas emissions and avoiding fumes and odors from the immediate surrounding areas, which represented a big challenge. The scientific community, several agencies and universities have been doing laboratory research of technologies that allow the reductions; these technologies are called warm mix asphalt (WMA). Subsequently some test sections were paved using the new technologies to ensure their success and gathering information of the performance of the pavement produced under those WMA conditions.

The WMA technologies evolved into different processes of production, the principal objective of this is to reduce the temperatures of mixing and compaction. The WMA design is just like the hot mix asphalt (HMA) design using the same practices with materials, proportions, volumetric parameters, mechanical properties, and performance test, among others.

The damage and performance of the pavements produced with WMA technologies are usually measured and compared with the performance of the HMA with parameters related to cracking, rutting, stripping, compaction level, moisture damage, modulus, etc.

Warm mix asphalt is a technology that is growing constantly. This chapter gives a brief description of WMA and presents the usage of a Green water-based additive to control and improve the characteristics of foamed asphalt.

2. Foamed asphalt with a Green water-based additive

2.1 The emerging of warm mix asphalt

The technology itself and its features emerged in Germany during the late 1900s. WMA became an important focus of research when the Kyoto protocol signed in 1997 entered into force in 2002. The same year the WMA term came for the first time in the United States of America (USA). In 2006 in the USA, the Federal Highway Administration and the National Asphalt Pavement Association (NAPA) formed a national WMA Technical Working Group [1]. Coupled with the environmental gains, additional economic and engineering benefits have motivated the improvement of methods to reduce the energy consumption at the asphalt pavement industry [2]. As a result, the developed WMA technologies allow the reduction of the temperatures at which commonly used HMA is typically produced, obtaining temperatures from 20 to 55°C lower [3].

2.2 Advantages and benefits

Since the WMA was conceived, several benefits had been consistently identified and consequently driving the development of new WMA technologies. Here they are presented [4–7]:

i. Environmental benefits: Less greenhouse gas emissions and fumes to the atmosphere due to the reduction in energy consumption. It is reported that the savings in burned fuel are around 20–35%.

ii. Paving benefits: The achieve workability facilitates greater hauling distances, an extended paving season, and faster opening to traffic.

iii. Economic benefits: In good operation conditions, the plant would have economic benefits through the reduced energy consumption.

iv. The welfare of workers: The exposure to fumes and odors narrows, since there is a correlation between production temperatures and asphalt fume in mix plant and field.

v. Versatility to produce the same gradation mixtures as the ones that are produced with HMA.

vi. Potential to incorporate higher percentages of reclaimed asphalt pavement (RAP) over 30% due to the increased workability of the mixture.

vii. Potential to place thick lifts and open to traffic in shorter times.

viii. Reduction in the asphalt aging due to the lower temperatures.

ix. Long lasting.

An important point to highlight is the possibility to increase the usage of RAP in WMA [8] due to the paving industry that is constantly facing the increasing demand for environmentally friendly pavement materials and the requirement of rising costs of raw materials. In connection with green technologies, the use of RAP represents a positive method of recycling and saving with the fact that in the

USA the asphalt pavements are the number one recycled material. The experience had shown that properly designed HMA containing RAP performs as well as HMA prepared exclusively with virgin materials. The increased workability of WMA represents potentially greater dosages of RAP. The improved workability leads to a lower production temperature and less aging of the binder, thus counteracting the stiffer RAP binder. Certain studies even recorded RAP percentages of over 50%. It has been shown that the RAP source and RAP content influence fuel consumption and emissions.

Many studies claim that the use of RAP in WMA pavements can help to offset initial costs, conserve natural resources, and avoid disposal problems. The properties of adequately designed recycled asphalt pavements have been proven to be comparable to all new materials asphalt concrete pavements.

2.3 Warm mix asphalt technologies

The WMA technologies are classified by type: organic additives, chemical additives, and those technologies that introduce water into the mix through a water-bearing additive or through a modification into the production process creating the foamed phenomenon. The organic and chemical additives diminish the friction by means of a reduction in the viscosity of the asphalt and by reducing the friction between the asphalt and aggregate, respectively. Foaming relies on the fact that when a given volume of water came in contact to over vaporization temperature-heated asphalt at atmospheric pressure, it expands into steam. When the water is dispersed in the heated asphalt and turns to steam, it results in an expansion of the binder phase and a corresponding reduction in the mix viscosity. The amount of expansion depends on a number of factors, including the amount of water added and the temperature of the binder.

2.3.1 Organic additives

This technology introduces organic additives and waxes to the mix, and when the temperature rises above the melting point of the waxes, there is usually a decrement in the viscosity. As the mixture cools, these additives solidify into microscopically small and uniformly distributed particles, which increase the stiffness of the binder in the same way as fiber-reinforced materials. It is recommended to take care of the selected type of wax in order to avoid possible temperature problems. Specifically, if the melting point of the wax is lower than in service temperatures, this can lead to complications.

Electing the right wax minimizes the embrittlement of the asphalt at low temperatures. The waxes in this technology are high-molecular hydrocarbon chains with a melting point of 80–120°C and are able to modify the properties of the original binder. The temperature at which the wax melts is in direct relation to the length of the carbon chain.

2.3.2 Chemical additives

In most of the cases, this type of additives is composed of a combination of emulsifier agents, surfactants, polymers, and additives to improve the coating and the mechanical characteristics of the mixtures. They reduce the friction generated between asphalt and aggregate particles making possible to produce the mixture at lower temperatures with better workability and compaction properties. Usually, these additives are mixed with the asphalt before it enters into the mixing drum.

2.3.3 Foamed asphalt

Mechanical foaming has become the most popular method to produce warm mix asphalt at least in the USA. The foaming caused by water-bearing additives includes inorganic synthetic crystals like zeolites. Zeolites gradually deliver their water at contact with heated asphalt producing the foaming effect. According to NAPA almost 65% of all WMA produced in 2017 were manufactured though foaming. The general mechanical process is based on introducing small amounts of cold water that are injected into a stream of the heated binder at temperatures ranging from 160 to 180°C. The process of mixing cold water with hot binder results in a volumetric expansion. Consequently, the viscosity of the binder is reduced which favors the coating of the aggregates and improves the mixture workability.

Although mechanical foaming has been a widely used technique in recent years, there are persisting questions regarding the effects of the water on binder foaming characteristics and foamed mixture properties and performance [9]. There are some parameters that can contribute to determining the efficiency of the process and the mixture.

Recently, at Texas University in the USA, a novel noncontact method to measure the expansion ratio and collapse of the foamed binder during the foaming process and the size and evolution of the number of foam bubbles over time has been developed. Two parameters were proposed for evaluating the effect of water content on foamed mixture properties in terms of workability and performance. The results indicated that the amount of water used in the foaming process had a significant effect on binder foaming characteristics and foamed mixture properties. The optimum foaming water content could be determined through a workability evaluation of foamed asphalt mixtures produced at different foaming contents.

2.3.4 Description of the equipment

The characteristics of the foamed binder are influence by factors such as binder source, water content, liquid additives, and the foaming properties of binders [10]. The foaming characteristics of the binder are then related to the coating and workability in order to estimate the optimum foamed water content and to validate the performance of the foamed mixture:

a. Laboratory foamer: a machine to produce foamed asphalt with pressurized air and water. In this foamer, it is possible to use different percentages of water to foam and has the capability to produce mix batches from 16 to 24 kg of aggregates. The pressured water and pressurized air are also variable.

b. Laser device and camera: the laser device consists of an emitter and detector laser to measure the distance from the device to the asphalt surface on the container; both are mounted on the triploids and aligned vertically and perpendicular to the ground. Both the laser and the camera are connected to a PC in order to obtain the data through time.

c. Pug mill: is a device for mixing the asphalt and the aggregates (16–24 kg).

See the equipment in **Figure 1**.

Some asphalts do not have the characteristics to be foamed. They lack a good expansion ratio and bubbles' average lifetime; in some cases there are special additives to promote foamability increasing those parameters (**Figure 2**).

Additionally, the use of digital images has improved the analysis of homogeneity of the bubbles. When the bubbles are analyzed at different times since they were

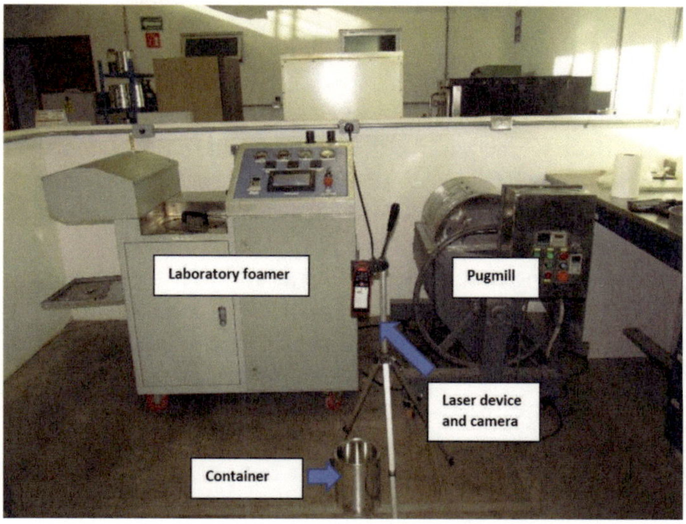

Figure 1.
Laboratory test setup for measuring binder foaming characteristics (Lasfalto Laboratory, 2018).

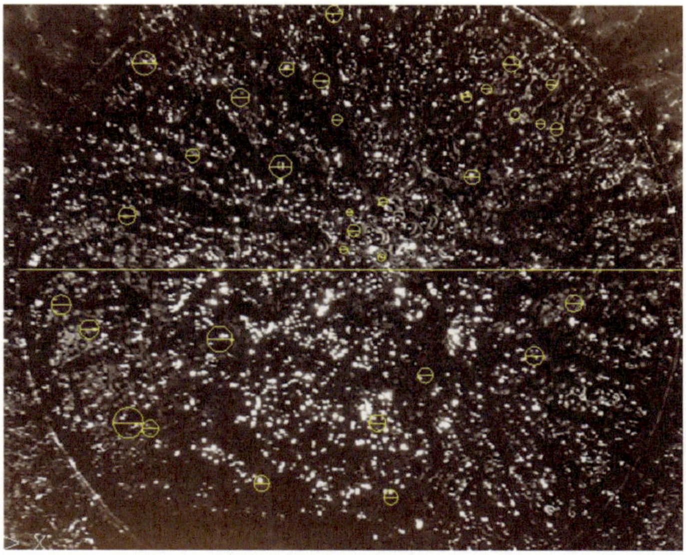

Figure 2.
Characteristics of a foam produced in the laboratory, in which bubbles with a homogeneous and stable size are observed (Lasfalto Laboratory, 2018).

formed, their evolution is observed in terms of expansion ratio, average half-life, and homogeneity. The target is to obtain homogeneous small bubbles, a reasonable expansion ratio, and an average half-life enough to coat the aggregate during the production process.

2.4 Warm mix asphalt with a Green water-based foaming additive

In this work, the performance of a Green water-based foam additive consisting of a package of neutral organic compounds, adhesion promoters, and water-based design to improve the characteristics of foamed asphalt was evaluated starting from the calculation of the ideal dosage. The results were compared to hot mix asphalt and foamed asphalt mixtures with no additives.

2.4.1 Experiment

During the experimental part of this work, a laboratory foamer described in Section 2.3 was utilized, and the characteristics of the foam were determined as explained above. Green water-based additive was used to promote the foaming effect and get the benefits of the WMA technology. The characteristics of the foamed asphalt obtained with the Green water-based additive were compared with the characteristics of foamed asphalt obtained just with water and no other additive. Both samples were foamed using the foamer machine, and the Green water-based additive was added in the water tank. A concentration of 2.1% of water and 2.1% of the Green water-based additive containing a 20% of solids was used. **Table 1** shows the obtained characteristics of the foamed asphalt.

The foamed asphalt specimens were prepared as follows:

i. The aggregates were dry and heated to a temperature between 130 and 140°C for WMA and 160°C for HMA.

ii. The Green water-based additive was injected into the foaming chamber, while the asphalt was at a temperature of 130°C for WMA and 160°C for HMA.

iii. The formed foam was mixed with the aggregate for 3 min at a temperature of 130°C for WMA and 160°C for HMA.

iv. Based on the NCHRO 807, the foamed WMA was aged at 240°F (115.5°C) for 2 h.

v. The mixtures were compacted at a Marshall compactor at a temperature of 110–120°C for WMA and 140–150°C for HMA.

The grain-sized distribution of the aggregates was taken by the granulometric curve of the regulation of the asphalt plant of the government of Mexico City. Marshall stability and flow test (ASTM D-1559), tensile strength ratio (TSR) (ASTM T-283), and resilient modulus (UNE-ENR697-26:2006) were performed to evaluate the mixtures.

The study of these parameters allows determining the resistance and durability of the asphalt mixture in the long term as a consequence of the production process.

Two series of three mixes were elaborated: foamed WMA with the Green water-based additive, foamed HMA and foamed WMA.

Table 2 shows the obtained results. The foamed HMA was used as a reference. Finally, the percentages of adhesion were calculated.

2.4.1.1 Results

The expansion coefficient increased to 50% when the foaming process was produced using the Green water-based additive compared to when the foaming process

Additive	Expansion coefficient	Half-life (s)	Average size (cm)
WMA (water)	4	18	0.16
Green water-based additive	6	70	1.61

Table 1.
Comparison of the characteristics of a water-foamed asphalt and a foamed asphalt using a Green water-based additive.

Additive	Asphalt (%)	Water or additive (%)	Resilient modulus (MPa)	Stability (kg)	Flow (mm)	TSR (%)	Adhesion (%)
HMA (water)	6	2.2	3800	>1200	<4	93	<50
WMA (water)	6	2.1	2108	>1000	<4	87	<50
WMA (Green water-based additive)	6	2.1	3420	>1000	<4	90	>86

Table 2.
Laboratory evaluation of the mixtures.

just used water, but the reduction in the average size diminished as much as 10 times using the Green water-based additive, and the half-life increased almost 4 times. These parameters indicate that the capability of the asphalt to coat the aggregate was substantially incremented because of the augmented contact area produced by the mass of smaller bubbles of asphalt. A notorious size homogeneity in the bubbles was observed and joined with the increased half-life of the bubbles it became in better homogeneity during the asphalt mixture process and prolonged workability of the mixture. In general, the characteristics of the foam are improved by the addition of the Green water-based additive, and these properties were expected to upgrade the properties of the warm mix asphalt elaborated with this foamed asphalt.

The obtaining results of the evaluation of the mixtures elaborated with foamed asphalt show that the resilient modulus and the stability of the WMA elaborated with the Green water-based additive hold a close behavior with a foamed HMA. The resilient modulus of the foamed HMA was 3500 MPa, and similar values were obtained to the mixtures prepared with the Green water-based additive with 3420 MPa. The resilient modulus for WMA foamed with water reaches the value of 2500 MPa which exhibits a less resistant mixture than the HMA and foamed WMA using the additive. In the Marshall test, we do not find significant differences because of the nature of the test. Nevertheless, in the test that indicates the humidity and water resistance that are the TSR values and the adhesion, the results show a desirable performance with the addition of the Green water-based additive reaching the value of 90% for TSR test and increasing the adhesion from <50 to >86%.

The aggregate characteristics made a mixture with the poor affinity between asphalt and aggregate. Even if it is recommended to have an adhesion percentage above 90%, the improvement in adhesion is an evidence of the functionality of the additive.

Mexico City can be considered as a pushful city in terms of selecting green technologies to maintain its paving needs. Since 2008 the asphalt plant belonging to the government of Mexico City and the one in charge of producing most of the asphalt mixture has been delivering WMA with an average production of 400,000 tons per year. Due to the high air pollution levels constantly registered in the urban area, one of the actions taken by the government was the mandatory use of WMA in every paving work at the city from 2010 to date. The registered reduction in fuel consumption has been approximately 20%, and, consequently, the VOC and the greenhouse gas emission have decreased. The Mexican plant had worked with several additives: some reduce the friction between aggregate particles and binder during mixing, and some are water-bearing additives. More recently, the company acquired an Astec Double Barrel facility with the so-called Green System, and the plant to improve the performance and the characteristics of the mixture is working with the Green water-based additive presented above.

3. Conclusions

Warm mix asphalt technologies have a great potential in different ways from improving the technologies to the implementation as a regular practice where asphalt mixtures are used. It is extensively recommended to promote this technology taking advantage of all its benefits.

The Green water-based additive showed numerous advantages. The additive increased the contact area of asphalt to achieve a total coating of the aggregate, forming a great amount of small homogeneous bubbles. Workability and lubricity into the asphalt mixture through the formation of microbubbles that results in better compaction to reach the required densities and stabilities were also improved. The performance of the obtained mixture was compared to a foamed HMA reaching similar resilient modulus values. The adhesion of the mixture was also enhanced.

Additionally, and in concordance with the other WMA technics, the Green water-based additive reduced the production of greenhouse gas emissions increasing the hauling and paving seasons and improving the working conditions.

Author details

Paula Cristina Arroyo-Martínez[1*], Rey Omar Adame-Hernández[1], Ignacio Roberto Cremades-Ibáñez[1], Israel Sandoval-Navarro[1] and Norma Angélica Sánchez-Flores[2]

1 Súrfax SA de CV, Guadalajara, Mexico

2 Applied Science and Technology Institute, National Autonomous University of Mexico, Mexico

*Address all correspondence to: paula@surfax.com.mx

References

[1] Buncher M. Warm Mix Asphalt technologies. (n.d.). Available from: http://asphaltmagazine.com/warm-mix-asphalt-technologies/ [Accessed: 2 February 2019]

[2] Rondón-Quintana HA, Hernández-Noguera JA, Reyes-Lizcano FA. A review of warm mix asphalt technology: Technical, economical and environmental aspects. Ingeniería e Investigación. 2015;**35**(3):5-18

[3] D'angelo J, Harm E, Bartoszek J, Baumgardner G, Corrigan M, Cowsert J, et al. Warm-Mix Asphalt: European Practice. Washington; 2008

[4] Taylor P, Kheradmand B, Muniandy R, Hua LT, Yunus RB. An overview of the emerging warm mix asphalt technology. International Journal of Pavement Engineering. 2014;**15**(1):79-94

[5] EDC-1: Warm Mix Asphalt | Federal Highway Administration, (n.d.). Available from: https://www.fhwa.dot.gov/innovation/everydaycounts/edc-1/wma.cfm [Accessed: 2 February, 2019]

[6] Chowdhury A, Button JW. A Review of Warm Mix Asphalt. Texas; 2008;**7**

[7] Rubio MC, Martínez G, Baena L, Moreno F. Warm mix asphalt: An overview. Journal of Cleaner Production. 2012;**24**:76-84. DOI: 10.1016/j.jclepro.2011.11.053

[8] West R, Willis JR, Marasteanu M. Improved Mix Design, Evaluation, and Materials Management Practices for Hot Mix Asphalt with High Reclaimed Asphalt Pavement Content. Washington; 2013

[9] Kumar GS, Suresha SN. State of the art review on mix design and mechanical properties of warm mix asphalt. Road Materials and Pavement Design. 2018:1-24. DOI: 10.1080/14680629.2018.1473284

[10] Arambula E, Newcomb DE. Effect of water content on binder foaming characteristics and foamed mixture properties. Transportation Research Record Journal of the Transportation. 2015:1-7 [Research Board 2506]

Chapter 3

The Effects of Using Nano-Silica in Cold-In Place Asphalt with Emulsified Bitumen

Shahab Hasaninasab

Abstract

Because hot mix asphalt (HMA) is both costly and harmful to the environment, the use of cold in-place recycled (CIR) asphalt is important to highway pavement. But, because CIR pavement properties are not as effective as HMA, addition of additive materials to the mixture is required to improve the pavement characteristics. Effects of using nano-materials on the mechanical properties of CIR asphalt are assessed in this chapter. The results show that the use of nano-materials improves the mechanical properties of CIR.

Keywords: cold in-place asphalt, nano-materials, emulsified asphalt

1. Introduction

The highway network of each country has a vital role in its economic growth and can determine the developed indexes. To address the high cost of implementing new roads, the losses resulting from damaged roads, or other multiple defects of conventional methods for road improvement, specialists have found new ways to improve traditional methods. Limited high-quality aggregate and resources have hindered the development and maintenance of traditional highway roads, making it harder to meet the needs of the society. Recycling is reusing of materials that already have their basic services. With the growing popularity of various recycling methods, asphalt recycling has also had significant growth.

The advantages to the approach presented in this chapter are wide range, all from economic benefits to conservation of natural resources. This method saves time and is in compliance with environmental requirements. It helps retain pavement thickness, rejuvenate and restore existing pavements, remove surface damages, and cure existing cracks. This method is also beneficial, as it does not require the discharge of waste materials by the depot, retains its quality after operation, maintains and develops safe traffic during the implementation of the project, reduces environmental pollution by preventing additional transportation, and applies new materials and sources of aggregates and bitumen.

To reuse the existing pavement, the reclaimed asphalt pavement (RAP) should be converted into usable materials. The well-graded RAP is a mixture of aggregate and good bitumen covered for asphalt production. The in-place recycling asphalt method is further broken down into cold in-place asphalt (CIA) and hot in-place recycling (HIR).

CIA is used in the processes of digging, plowing, scraping, and crushing asphalt layer materials. It is also used in a central asphalt factory to reprocess with bituminous materials at room temperature and without the use of heat. The stratum of recycling is usually coated in proportion to the physical condition with asphalt concrete or surface asphalt [1–4].

Emulsified asphalt pavement has initial and low Marshall stability. The use of additives causes resistance in the produced Marshall asphalt at an early age or later, which can also be effective in improving cold asphalt resistance. In the case of using cement in cold-recycled asphalt to achieve a certain level of density, less energy is required, and the use of emulsified asphalt allows for a more uniform and consistent asphalt. Miró Recasens et al. found that the emulsified asphalt that has cement with Marshall resistance has higher resilience than that without cement. This is mainly due to the fact that higher flexibility, cracking, and the potential of increasing the crack in emulsion asphalt with cement are less [5].

It was observed that in addition to the earlier mentioned benefits of using cement in cold-recycled asphalt, the use of adequate cement in cold asphalt mixture could greatly reduce harmful effects, such as asphalt creeps and cracks. Researchers have found the mathematical relationship between the elasticity model and the amount of used cement in the cold asphalt mixture, along with its preparation process [6].

The effect of modified and unmodified nano-clay in the production of hot asphalt, the used nano-clay from the montmorillonite, has been investigated. In this study, a small percentage of nano-material was observed to have had a great impact on asphalt properties [7–8]. This study also found that a 2% addition of nano-clay to binder asphalt increases the combined shear model to more than 184%. For hot asphalt, modified polymer with nano-clay was used in this study to improve the mechanical properties and thermal resistance of the asphalt. The modification reduces asphalt permeability while improving rupture and compressive strength. Using the modified nano-clay in asphalt causes reduced asphalt viscosity, making it easier to implement the asphalt. The use of modified nano-clay also increases the ability of asphalt recycling. Microscopic photos illustrate how the use of modified nano-clay can lead to better and more complete molecular bonds. The use of nanoparticles increases durability of the produced asphalt. For example, with the use of 2–4% of nanoparticles in asphalt, the crack depth is half reduced [11].

In their article “The use of nano-silica in concrete mixes,” Garcia and Brawerz of Eindhoven University in the Netherlands indicated that the use of silica in place of cement reduces the cement-produced CO_2 level and increases the strength and durability of the concrete, due mainly to the use of nano-silica in the concrete mix [9, 10].

The trial, however, did not focus on the effects of nanoparticles on cold-recycled asphalt. This chapter details the financial, environmental, and other factors (not including hot and emulsified asphalt) affecting the performance and development of civil engineering projects that use cold-recycled asphalt. This chapter investigates the effect of nano-materials on the recycling cold mixture.

2. Methodology

The modified Marshall method, which is based on ASTM-D1559 or AASHTO T245 and is different from hot mixed asphalt (HMA), is used for CIA with emulsified asphalt. The emulsified asphalt in this design is used in mixtures that are prepared in accordance with ASTM-D5505; RAP materials, and if necessary, new aggregate materials are used as well. The amount of used emulsion in the Marshall mix method ranges usually from 1 to 2%, and young emulsions vary from 5.0 to 1.25% compared to the weight of the asphalt mixes. The more the mixtures, the higher their percent of broken aggregates, and a lesser amount of available bitumen is used in RAP. Lower percentages are used in mixes with high levels of fine and rounded corners. When the new aggregate is added to RAP, the young emulsion is not used, while the percentage of emulsified asphalt should be increased. In order to estimate the percentage of the primary emulsified asphalt for this design, a method has been used in which a given amount of emulsified asphalt was chosen as the base percentage; and the amount of bitumen and the degree of recovered bitumen penetration from hard bitumen, in terms of RAP grading, are reformed.

Nanoparticles are defined as particles smaller than 0.09 mm. Some methods of producing nanoparticles include grinding through washing or replacing. These methods often require advanced nuclear devices.

Properties of nano-material are always defined in nano-material mass, namely, the properties of the particles themselves are not apart from the properties of the particle bulk, and the behavior of the particle bulk represents the mechanical and chemical properties of nanoparticles.

Since each nanoparticle has a unique size, shape, and surface, their behavior and reaction should be examined in combination with other available substances [11].

The following formula is used to estimate the amount of needed emulsified asphalt for the recycled mixture:

$$E = 1.2 + A_G + A_{AC} + A_{PV} \quad (1)$$

where, E = the percentage of required bitumen, 1.2 = the fixed percentage of bitumen, A_G = the correction factor for the size in terms of percent, A_{AC} = the correction factor for the percentage of existing bitumen in recycled mixture in terms of percent, A_{PV} = the correction factor for penetration grade.

In cases where E comes to have different results for a recycled material sample, the bitumen percentage is shown to be less (Publication 339). The water amount can be obtained from the following equation:

The water mixture should be added = 3-(the moisture percent of aggregate mixture + the percentage of the water in bitumen mixture). Nano-material has been added to the water and mixed; then the blended materials have been added to the emulsified asphalt.

Emulsified asphalt, which has been heated to 60°C, is added to the mixture, in accordance with the project's needs for each specific bitumen percentage and 0.5 difference. The resulting mixture is stirred until the emulsified asphalt or rejuvenating emulsion is evenly distributed in the material.

In order to detect changes in the percentage of additives to the asphalt, three compacted Marshall samples (by the standards of ASTM-D1559 and AASHTO-T254); one crushed sample, for determining the real specific weight (by the standards of ASTM-D2726 and AASHTO-T166); and one non-crushed sample, for determining the maximum real density and the percentage of the sample's free space (by the standards ASTM-D2041 and AASHTO-T209), were created.

The initial speculation for the addition of the emulsion to the mix is as follows:

Percentage of bitumen = 1.2% + 0% + 0% + 2% + 0% =

According to the above speculation, the three Marshall sample series with values of 1.5, 2, and 2.5% of emulsified asphalt were made without the addition of any additives. Since adding 3% of water to the produced mixture can help make a better blend of the mixture, the samples in this project were combined with 3% of water to the cold asphalt mixture and then tested.

To further examine the compatibility of cement and nano-silica, nano-alumina, nano-clay, and nano-lime with cationic emulsion and nano-silica and cement with each other (both have a negative surface charge), the combinations were used and tested for the first time in recycled mixture of cold asphalt. Three sets of samples with values of 0.5, 1, and 1.5% of nano-silica were obtained to determine the optimal amount of nano-silica when combined with optimal cement. **Figures 1–3** show the effects of nanoparticles in fatigue property of CIR.

The different parameters and additives were considered for each case as the obtained results were compared with the results of previous studies and the conformance or nonconformance results.

All nano-silica, nano-alumina, nano-clay, and nano-lime and cement were combined and used in this study, first because of their compatibility in cationic emulsified asphalt and second because of the effects cement has on increasing the Marshall stability of CIP.

The combined use of nanoparticle and cement are effective in increasing the ultimate strength and in protecting the environment due to the effects combined nanoparticle and cement have on increasing the Marshall stability of the produced samples and reducing the level of the produced CO_2.

The higher the real specific weight, the more rigid the produced pavement is. As resistance against pavement fatigue is improved, the resistance and the reliability of the pavement are increased as well. The actual real specific weights of the produced samples, compared to samples produced without additives, are increased by 8 and 10% with the addition of cement and nano-silica, respectively.

The addition of nano-silica and cement at the same time increased the strength and resistance to pavement fatigue and reduced parts of the asphalt flow and the comfort while the car is moving on the pavement. The addition of any additives to the cold asphalt mixture reduces the sample's free space percentage and results in a more compacted and condensed body. When nano-silica and cement are combined, the reduction in free space is impressive.

The combination increases the sample's air void by 20% without the additive. The additives can help limit the influence of surface water from the pavement to the body with regard to the provision of proper drainage in the lower layers of the pavement.

On the other hand, the presence of other additional fillers (additives) fills the aggregate pores with more fine materials, making the mixture more efficient. Finally, in addition to the positive impacts of the earlier stated factors, the effects were observed in the increased Marshall stability and resistance against fatigue in the cold asphalt mixture.

Other parameters of the cold asphalt mixture and the potential to replace this type of asphalt with hot mixed asphalt were considered by taking advantage of CIP instead of HMA to examine the measure of flow and the existing limitations.

The flow of samples made with nano-silica and cement additives was acceptable. Due to the maximum real density of the samples, it can be stated that with the addition of various additives, the maximum theoretical density of asphalt is higher.

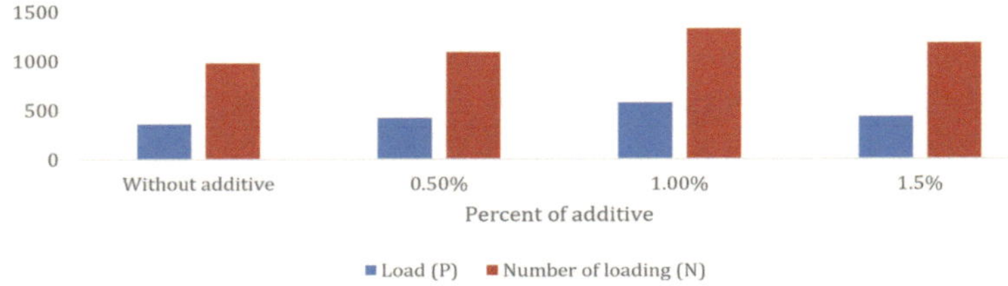

Figure 1.
Comparison of load and number of repetitions in stress ratio 0.2.

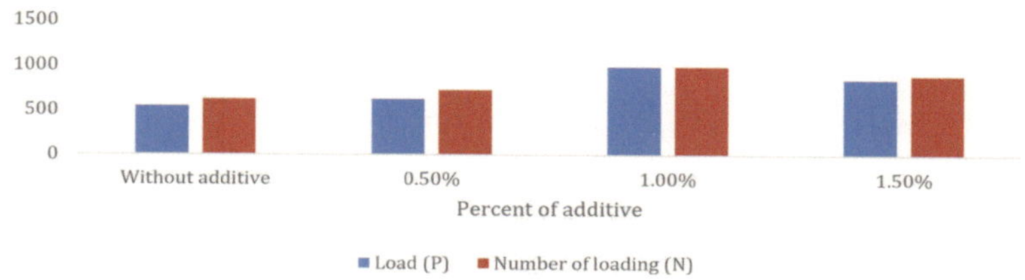

Figure 2.
Comparison of load and number of repetitions in stress ratio.

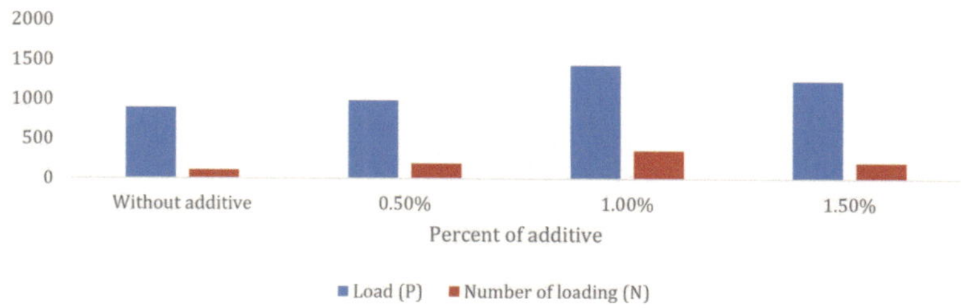

Figure 3.
Comparison of load and number of repetitions in stress ratio 0.5.

Active fillers, which have higher adhesion properties, cause the specific weight gain because of the physical and chemical bonds between aggregates. The addition of nano-silica and cement has improved the maximum theoretical density of the asphalt mixture by 3 and 4% compared to the mixture without additives. The increase in maximum theoretical density of additives with more adhesion strength is more; these findings are consistent with the assumptions made at the initial stages of this research.

3. Conclusion

Cold emulsified asphalt has low initial value and Marshall resistance. Additives can, therefore, be used to improve the mechanical properties. According to the assumptions made for this research, the use of nanoparticle rather than an active filler in a recycled cold asphalt mixture improves the Marshall resistance of samples and reduces the free space percentage of the mixture. It is anticipated that the use of nanoparticles in place of the active filler improves efficiency of the mixture and homogenizes the asphalt mixture. This study attempted to find a logical relationship between the types of additives in recycled cold asphalt and the extent of effective parameter changes made in asphalt while also commenting about the positive and negative effects of each additive on the different parameters of recycled cold asphalt.

Author details

Shahab Hasaninasab
Civil Engineering Department, Highway and Transportation Engineering, Faculty of Engineering, Razi University, Islamic Republic of Iran

*Address all correspondence to: s.hasani@razi.ac.ir

References

[1] Asi IM. Performance evaluation of Superpave and Marshall asphalt mix design to suit Jordan climatic and traffic condition. Construction and Building Materials. 2007;**21**(8):1732-1740

[2] Little DN, Petersen CJ. Unique effects of hydrated lime filler on the performance-related properties of asphalt cements: Physical and chemical interactions revisited. Journal of Materials in Civil Engineering. 2005;**17**(2)

[3] Rastegar M, Kokabi M. The effect of nano-clay on the morphology of polymer/bitumen blend. In: 1st International Bitumen Conference; Tehran, Iran; 2008

[4] Seref O, Fazil C, Atakan A. Performance of cement modified dense graded cold-mix asphalt and establishing mathematical model. Indian Journal of Engineering and Materials Sciences. 2007;**13**(6):512-519

[5] Recasens RM, Jiménez FEP, Aguilar SC. Mixed recycling with emulsion and cement of asphalt pavements. Design procedure and improvements achieved. Materials and Structure. 2007;**33**(5):324-330

[6] Cross S, Kearney A, Edward R, Justus HG, Chesner WH. Cold-In-Place Recycling in New York State. 2010

[7] Gallivan V. Cold in-Place Recycling State of Practice Review. 2011

[8] Thomas T. Mix and Thickness Design for Cold In-place Recycling (CIR). 2012

[9] Marshall RT, Garcia L, Carpenter SH. Cold In-place Recycling and Full-Depth Recycling with Asphalt Products (CIR & FDRWAP). 2009

[10] Quercia G, Brouwers HGH. Application of Nano Silica (NS) in Concrete Mixtures. At Lyngby, Denmark: 8th fib PhD Symposium in Civil Engineering; 2009

[11] You Z. Nanomaterials in asphalt pavements. International Journal of Pavement Research and Technology. January 2013;**6**(3)

Chapter 4

Incorporating Sustainable Practices in Asphalt Industry

Mohammadsoroush Tafazzoli

Abstract

Shrink of nonrenewable natural resources and the pollution generated by many manufacturing industries have initiated a global determination for pushing the industry toward more sustainable products. Asphalt as the principal element of almost any street and highway pavement is integral in transportation development, which, in turn, is pivotal for sustainable development. On the other hand, the material consumption and pollution generated in the asphalt industry pose significant threats to the environment and, therefore, to sustainability. This chapter reflects some of the sustainability concerns of the asphalt industry and discusses some of the possible solutions to mitigate them. The sustainability considerations are categorized into four phases of asphalt life cycle namely: (1) extraction of materials, (2) processing of raw materials, (3) transportation of asphalt raw materials and products, and (4) reusing and recycling at the end of a pavement's life. In each phase, best practices to improve and maintain the sustainability of asphalt pavements are discussed. This chapter also discusses sustainable approaches and technologies in the asphalt industry such as warm mixes, permeable asphalt pavements as well as the potentials for improving the mechanical properties of asphalt pavements particularly in terms of resisting heavy load traffic, clogging, and freeze-thaw.

Keywords: asphalt pavement sustainability, warm mix, permeability, emission

1. Introduction

The massive demand for asphalt pavement is a response to the growing length of highways and the number of vehicles worldwide. In the USA, for example, highway networks are approximately 8 million miles and facilitate over 3 million vehicle miles yearly [1]. According to Green Car Reports, there are nearly 1.2 billion vehicles in operation today, and this number is constantly growing to reach the estimated number of 2.5 billion by 2050. The massive number of vehicles has expanded the need for more streets and highways most of which require asphalt paving. Additionally, the pavements that are in use require periodic repairs and replacements of asphalt. This explains an enormous increasing demand for asphalt.

This chapter discusses a comprehensive approach to maximize the sustainability of the asphalt industry. This approach is presented in the two following major parts. (1) Assessment of asphalt's environmental impacts in its life cycle. In this section, the sustainable practices in the four phases of asphalt's life cycle (material procurement, processing of raw materials, transportation of products, and end of a product's life) are discussed. (2) Sustainable approaches in the asphalt industry. This section is divided into two parts that discuss two main directions in pushing the asphalt industry to improve its contribution to sustainability. The first direction is enhancing the technology of warm mix asphalts, which explains the benefits and the existing shortcomings of these mixes that require further research. The second part discusses the opportunities for reducing the transfer of pollutants through runoffs by using permeable asphalt pavements. The barriers to more widespread use of permeable asphalt pavements that need to be considered by the asphalt industry are also presented.

The chapter is expected to provide an understanding of the adverse impacts of the asphalt industry on the environment, how these impacts can be mitigated, and what additional steps must be taken to enhance the existing sustainable methods in different stages of asphalt's life cycle.

2. Is asphalt an unsustainable product by itself?

Despite the impression from the detrimental impacts of asphalt on the environment, in compassion with other pavements, asphalt can be addressed as a sustainable material. According to Gambetes, producing asphalt pavements requires almost 20% less energy than other pavements. Additionally, the construction and rehabilitation of these pavements are relatively quick as no curing is needed before the repaired road segment is ready to use. This is beneficial in reducing the exposure to construction process pollutions, in saving taxpayers' money, and in reducing the time of road closure, which contributes to the societal bottom line of suitability.

Recycling in the asphalt industry is not a goal for the future. It is already happening in massive amounts. For example, the USA, as the leading recycler of asphalt, reclaims approximately 65 million tons of its products annually and reused almost 99% of it [2]. The asphalt industry also contributes to the reuse of other materials such as tires, glass, blast furnace slag, and asphalt roofing shingles.

In terms of maintenance and rehabilitation, asphalt pavements are also less expensive than concrete pavements. Although their service life is relatively shorter compared with concrete pavements, with proper design, construction, and maintenance, they can continue to serve for decades without wearing out. Their maintenance is faster and less expensive than concrete pavements. This is why rubbishing or covering worn out concrete pavements with an asphalt surface is a common practice for repairing concrete pavements.

Research findings support that asphalt pavements provide superior contact between the vehicle tires and the road surface, which enhances vehicle safety. When designed with open-graded, asphalt pavements can effectively reduce the rainwater splash, which can potentially improve visibility and reduce accidents.

Asphalt pavements are relatively low noise pavements and are beneficial to be used in urban areas where reducing noise pollution is essential. According to the research [2], asphalt pavements can lower the noise between 3 and 10 dB, which affects similar to doubling the distance from the road to the people who can hear the road noise (**Figure 1**).

To answer the question if asphalt pavements contribute to sustainability or are detrimental to it, certain facts must be considered as follows:

1. The sustainability must be compared with other alternatives of pavements. If a pavement is more sustainable than other alternatives, even if, it has a burden on the sustainability triple bottom lines, it is considered as the sustainable

Figure 1.
The sustainability benefits of asphalt pavements compared with other pavements.

option. Based on this comparison, asphalt pavements are one of the most sustainable options for pavements.

2. The sustainability must be compared with the optimum potential sustainability that can be achieved in the procurement, design, production, shipping, construction, rehabilitation, replacement, and reuse of the product. Based on this approach, there is high potential in modifying the technology of asphalt pavements to enhance their sustainability.

3. How does the asphalt industry affect the environment?

The asphalt industry negatively impacts the environment in three ways. This is shown in **Figure 2**. As a petroleum-based product, asphalt mixes carry a significant amount of chemicals that can impact the environment during extraction, transportation of raw materials, processing, and using.

Another primary adverse environmental impact of asphalt is on water. Hazardous chemicals, dust, and other particles can penetrate to asphalt paving surfaces and then be transferred through stormwater runoffs. The polluted runoffs can eventually affect water bodies and endanger aquatic life, reduce water quality, and degrade the views to name a few. This explains the asphalt industry must focus vigorously on the ways to mitigate the accumulation and transfer of pollutants. The third major adverse impact of the asphalt industry is generating atmospheric emissions. A wide range of hazardous emissions are released particularly in the process of heating the mix. These emissions contribute to global warming and lead to various impacts on the environment as well as on the human health.

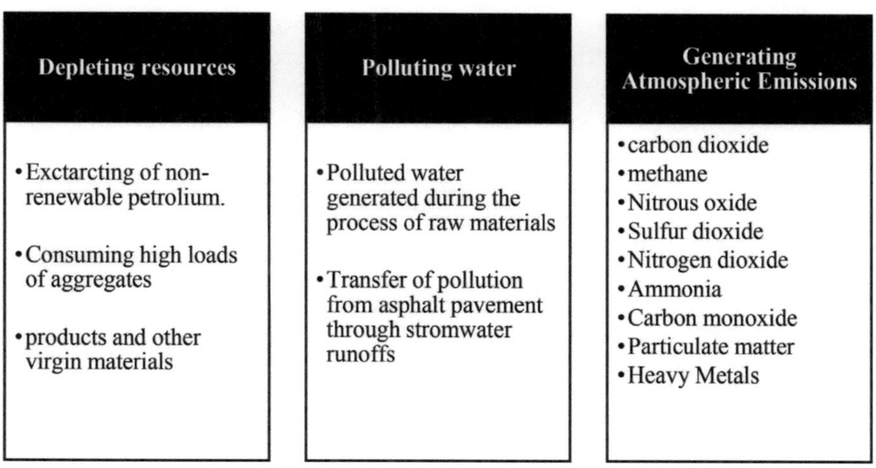

Figure 2.
The primary adverse impacts of the asphalt industry on the planet.

4. What are the main approaches in improving the asphalt industry's sustainability?

Once the detrimental impacts of asphalt on the environment are identified, practices to minimize these impacts must be established and followed. **Figure 3** shows some of the indicators of asphalt sustainability as it relates to material, construction, maintenance, and use [3]. The figure indicates why the sustainability of asphalt industry has significant potentials for improvement in multiple aspects. The challenge is the tradeoff of compromising some of the mechanical properties of asphalt to make it more sustainable. For instance, a wide range of emissions can be reduced by lowering the temperature of heating the asphalt mixes, but there are some uncertainties if the reduced temperature can maintain an asphalt pavements resistance against fatigue or heavy loads. This will be discussed in more details in Section 5.2.2. Therefore, the sustainable approaches must keep a balance between lowering the adverse environmental impacts and the functionality, mechanical property, and durability of asphalt pavements.

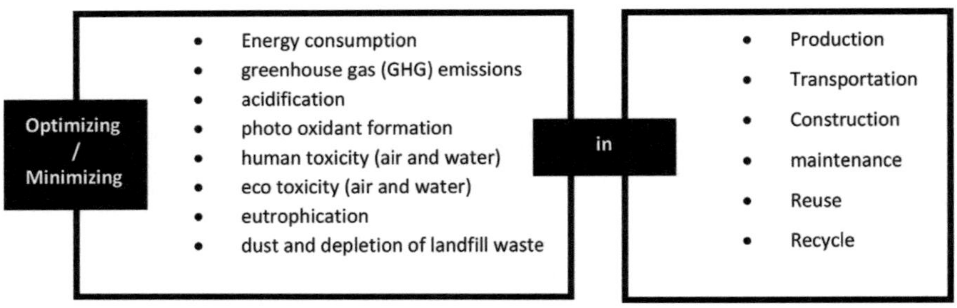

Figure 3.
Potential approaches to improve environmental sustainability in the asphalt industry.

5. Mitigating asphalt's negative environmental impacts in its life cycle

The life cycle assessment or cradle-to-grave is an integral element of sustainability evaluation. This approach must be applied for asphalt analysis to assess its impacts on global warming and climate change, depletion of fossil fuels, and human well-being to name a few [4]. Life cycle assessment encompasses a wide range of decisions and activities in different phases of asphalt production, transportation, use, and reuse. In addition to extraction, significant amounts of materials such as bitumen, fuel oil, and gasoline are transferred and stored in asphalt plants. Opportunities to modify the traditional practices that are commonly used in asphalt plants must be studied for each plant. There exist common best practices that can be followed by the managers of asphalt plants. **Table 1** shows the four primary steps in life assessment of asphalt products (data from [5]). These four stages and some best practices to apply in each of them are discussed in the following.

5.1 Material procurement (phase 1)

The practices used in the procurement of virgin materials can significantly identify the level of sustainable production of asphalt mixes. While main asphalt ingredients are primarily byproducts of the petroleum industry; considering the magnitude of their production, there are independent practices for enhancing the sustainable procurement of virgin materials in the asphalt industry. Based on the discussed sustainability concerns about the asphalt industry, the following target can be defined to push this industry toward more sustainable practices:

5.1.1 Extraction of materials

The enormous demand for nonrenewable raw materials to produce asphalt pavements is a threat to sustainability. For instance, 83% of all street pavements in the USA are made of asphalt [6]. The primary step in contributing to sustainability in the procurement of asphalt materials is reducing the demand for the extraction of nonrenewable petroleum resources and other virgin materials. One of the greatest advantages of asphalt is its recyclability for almost unlimited times. The fact that asphalt pavements are 100% recyclable can be interpreted that there is no certain end life for them [7]. **Figure 4** shows the sustainable procurement process of

Cradle to grave process	Phases	How the assessment must be performed
Cradle to gate	Phase 1. Material procurement	• Extraction of materials • Reuse and recycle
	Phase 2. Processing of raw materials	• Energy consumption reduction • Pollution and emission reduction
	Phase 3. Transportation of product	• Vehicles' air pollution reduction • Preventing the spill of polluting materials during transportation
Gate to grave	Phase 4. End of a product's life	• Connecting end use point to reuse and recycle

Table 1.
The primary steps of a life-cycle assessment of asphalt products.

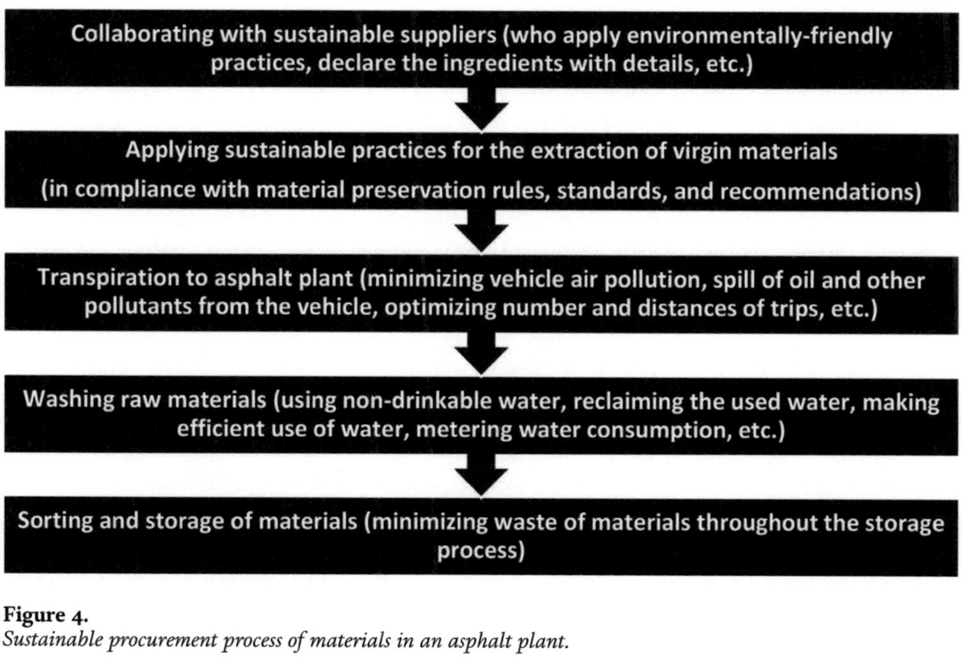

Figure 4.
Sustainable procurement process of materials in an asphalt plant.

materials in an asphalt plant. As it is seen in each of the six following steps, there are potentials to maintain the sustainability of operations. The asphalt plant's determination and leadership in establishing these practices is pivotal in the successful application of these practices and policies.

5.2 Processing of raw materials (phase 2)

The most energy-intensive phase of asphalt production is the processing phase as it requires a high temperature for burning and processing of raw materials. This phase is also the most polluting phase due to the emissions that are normally generated in different heating processes. Therefore, energy and pollution reduction are the primary sustainability goals of this phase. This is explained in the following.

5.2.1 Energy consumption reduction

To have a holistic energy consumption reduction approach, the consumers of energy in the typical production phase should first be identified. In the following, some of the main energy consumers in the production process are identified. (1) Fuel or electricity needed for construction equipment that is applied in the production process, (2) energy to operate burners that dry aggregates, and (3) fuel for heating liquid asphalt binder and other mixing processes (natural gas, propane, diesel fuel, recycled fuel, biofuels, etc.)

The comprehensive study of the system to identify the potentials for energy consumption reduction is required. Some of the best practices in this regard are the following:

1. Supplying energy from more sustainable sources (solar and wind power)

2. Constant maintenance of the tools and equipment to maintain their efficiency

3. Minimizing wastes of energy in heating processes

4. Replacing low-efficiency tools with new efficient ones

5. Optimizing the layout, production queuing, and supply of raw materials to minimize energy consumption

5.2.2 Pollution and emission reduction

The second main goal in sustainable processing of asphalt is pollution and emission reduction. To enhance the environmental sustainability of the asphalt industry, controlling certain chemicals must be the target. **Table 2** shows the primary hazardous chemicals, how they are generated in the asphalt industry, and why they must be controlled.

It should be noted that lowering the mixing temperature can substantially contribute to both energy consumption and emission reduction. Therefore, it is crucial to invest in modifying asphalt technology in this direction. Warm mix asphalt is the product of this approach, which is one of the most impactful modifications in asphalt technology to make it more sustainable. Warm mixes will be discussed in Section 6.1.

Research shows that the reduced energy consumption required in the manufacture of warm mix asphalts can lower the carbon dioxide emission between 30 to 40% [8] and 30 to 50% reduction on the fumes at bituminous plants, which is

Hazardous chemicals	Why they must be controlled (selected hazards and the chemical causes)	Primary generation area
Carbon dioxide (CO_2)	Global warming, climate change, human toxicity	Construction
Methane (CH_4)	Global warming (21 times more than carbon dioxide), photo-oxidant formation	Construction
Nitrous oxide (N_2O)	Global warming	Construction
Sulfur dioxide (SO_2)	Acidification, photo-oxidant formation, human toxicity	Construction
Nitrogen dioxide (NO_2)	Acidification, eutrophication, human toxicity, photo-oxidant formation	Construction + End of life
Ammonia (NH_3)	Acidification, eutrophication, human toxicity	Construction + end of life
Carbon monoxide (CO)	Human toxicity, photo-oxidant formation	Construction
Nonmethane volatile organic compound (NMVOC)	Ecotoxicity, human toxicity, photo-oxidant formation	Construction + maintenance
Hydrocarbon (HC)	Ecotoxicity, human toxicity	Construction + maintenance + use
Particulate matter (PM)	Human toxicity	Construction
Heavy metals	Human toxicity	Construction
Arsenic (As)	Ecotoxicity, human toxicity	Maintenance + use
Cadmium (Cd)	Ecotoxicity, human toxicity	Maintenance + use
Lead (Pb)	Ecotoxicity, human toxicity	Maintenance + use
Mercury (Hg)	Ecotoxicity, human toxicity	Maintenance + use
Chemical oxygen demand (COD)	Eutrophication	End of life
Phosphate (PO_4)	Eutrophication	End of life
Nitrate	Eutrophication	End of life
Phosphorus (P)	Eutrophication	End of life
Nitrogen (N)	Eutrophication	End of life

Table 2.
Hazardous chemicals that must be controlled to enhance the environmental sustainability of the asphalt industry.

significantly beneficial for workers on a paving team [9]. Additionally, reduced emissions enable to establish asphalt plants closer to urban areas [10]. This, in turn, can reduce transportation pollution, cost, and increase the accessibility of asphalt materials for urban construction. **Table 3** compares the emission generated by hot and warm mix asphalts ([11] as cited by [8]).

As it is seen in **Table 3**, the amount of emission reductions for replacing hot mix with warm mix asphalts is significant. These reductions contribute to reducing the global warming effects at a large scale, pollution for neighboring urban areas at a middle scale, and a healthier environment for the people working at the asphalt plant in a small scale.

5.3 Transportation of asphalt raw materials and products (phase 3)

Asphalt industry is involved with a massive demand for transportation to (1) supply the raw materials for the production and (2) deliver the products to consumers in an almost nonstop basis. Additionally, being a petroleum product, asphalt materials and products carry high loads of chemicals that can potentially be hazardous to the environment (as it was listed in the **Table 3**). **Table 4** shows some of the indicators of sustainable transportation of asphalt raw materials and products.

5.4 End of a product's life (phase 4)

Considering the huge amount of asphalt-paved streets and highways, an enormous amount of asphalt pavements can reach their end-of-life point in any selected interval. The sustainable approach of cradle to cradle attempts to connect the end of

Type of emission	Hot mix asphalt	Warm mix asphalt	Reduction
Temperatures (°C)	155–165	110–120	45
Gases temperature (°C)	65.6	50.3	15
CO_2 (%)	2.12	1.59	25
CO (ppm)	217	151.6	30
NO_x (mg/m^3, e.g., NO_2)	26.8	21.5	25
Environmental dust (mg/m^3)	168	21	88

Table 3.
Comparing the emission generated by hot and warm mix asphalt.

Sustainable transportation approach	How to implement
Reducing the number of trips	• Increase the capacity of delivery in each trip • Optimize trips to deliver more in fewer trips
Enhancing the safety of trips	• Building private roads for main access to the asphalt plant • Putting signs around the heavy vehicle
Reducing the distance of trips	• Locating the asphalt plants with optimal distance to multiple consumers • Creating temporary asphalt plants near road projects
Protecting health and well-being	• Reduce the vehicles' pollution or use hybrid vehicles • Reduce transportation noise
Protecting the hosting roads environment	• Minimize spill of chemicals form the vehicle • Be responsible for fixing possible damages to the environment • Ongoing maintenance and smog check of vehicles • Replacing polluting and old vehicles with efficient and green vehicles
Promote local consumption	• Give priority to local projects to avoid long trips

Table 4.
Principles of sustainable transportation of asphalt raw materials and products.

life to reuse of asphalt pavements. The primary solution to reduce the demand for more raw materials for asphalt production is the reuse of asphalt and mixing it with virgin materials. Asphalt mixes contain a considerable amount of recyclable materials such as coarse and fine aggregates. To maximize the asphalt reuse, production of certain types of asphalts must be encouraged.

5.4.1 Reclaimed asphalt pavements (RAP)

Using reclaimed asphalt pavement (RAP) in producing new bituminous mixes has multiple sustainability advantages including significant cost and energy savings and environmental benefits due to a reduction in the waste generated in road maintenance and rehabilitation processes [12, 13].

According to the National Asphalt Pavement Association of the USA, reclaimed asphalt pavements are removed or reprocessed pavement materials containing asphalt and aggregates. These materials are the debris of construction activities such as reconstruction, resurfacing, and diggings to gain access to buried utilities [14]. Based on the estimates, 41 million metric tons of RAP is produced yearly in the United States [14]. According to EAPA [15], in Europe, approximately 50 million tons of RAP is produced annually. This enormous volume of RAP can be reused to produce new bituminous mixtures [8].

The environmental and economic advantages of applying RAP in hot mix asphalts can be maximized by producing totally recycled hot mixed produced with 100% RAP. The challenge to do this is improving the mechanical properties of these mixes such as workability, durability, binder aging, and fatigue cracking resistance [16].

5.4.2 Recycled asphalt shingles (RAS)

5.4.2.1 Waste plastic

Waste plastic is another massive source of material reuse to procure the asphalt ingredients. Only in the USA, 42.6 1-l bottles are purchased every year. It is estimated that on a global scale, 200 billion plastic water bottles are consumed by people [17]. The abundant amount of plastic waste and its suitability to be used in the asphalt industry creates a massive opportunity to reduce the need for virgin materials. The research has identified and encouraged this opportunity [18, 19].

6. Sustainable approaches in the asphalt industry

This section discusses the next steps in improving the sustainability of asphalt industry. The two main approaches discussed are (1) warm mix asphalts and (2) permeable asphalt pavements. In each section, the sustainability advantages are explained. Additionally, the shortcomings of the existing technologies and potentials for improvement are discussed. **Table 5** provides a summary of this section.

6.1 Warm mix asphalts

Reducing the temperature of mixes is a fundamental sustainability approach in asphalt industry as it contributes to both energy reduction and emission mitigation [9, 10, 20, 21]. Warm mix asphalt is the result of the temperature reduction approach. Research findings support that using them contributes to reducing energy consumption and air emissions [22]. Regular hot mixes need a temperature of 150°C or more for production and compaction [23]. Warm mixes can lower the demand for energy by reducing the heat needed between 16 and 55°C (**Figure 5**) [24].

The decline of required temperature for processing asphalt mixes is proportional with requiring less fuel/ton to heat the mix. For instance, hot mix asphalt may require a temperature of 180°C and more than 7 l of fuel per ton and half warm asphalt may require to be heated between 60 and 80°C and less than 4 l of fuel per ton [9]. Reduction of fuel, in turn, leads to a reduction in fuel costs between

Technology type	How it contributes to sustainability	Shortcomings that need to be resolved
Warm mix asphalts	• Reduces energy demand for heating the mix • Reduces emissions	• Low resistance to fatigue • Moisture damage due to incomplete drying of aggregates
Permeable asphalt pavements	• Controls urban runoffs • Prevents transfer of pollution to water bodies	• Low resistance to high compressive loads • Clogging • Freeze-thaw issues

Table 5.
Sustainable technologies in asphalt industry.

Figure 5.
Low temperature of production for warm mix asphalt (courtesy of Federal Highway administration).

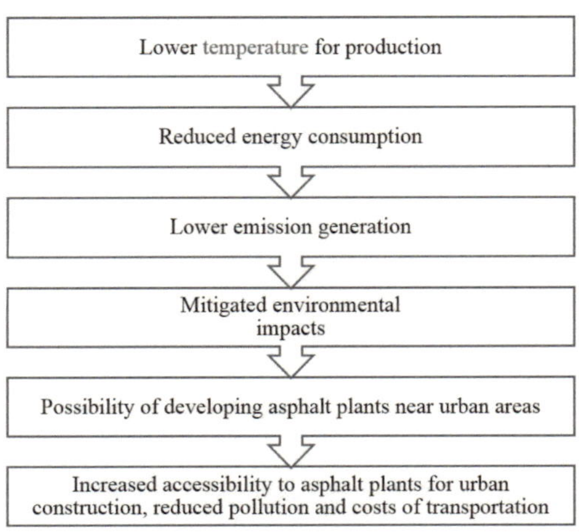

Figure 6.
How warm mixes contribute to sustainability.

25 and 35% [11]. **Figure 6** shows some of the sustainability advantages of these mixes (data from [25]).

Another advantage of warm mixes is the possibility of using reclaimed asphalt pavements in them [8]. In addition, sustainability advantages (see **Figure 7**) and research findings indicate that the mechanical properties of these mixes are not compromised to make them more environmentally friendly. These mixes have stiffness very close to hot mixes, and their water sensitivity and fatigue resistance are comparable with hot mixes.

6.1.1 Research status of warm mix asphalt mixtures

An impactful approach in increasing the sustainability of warm mix asphalts is the use of rubberized asphalt concrete (RAC). This novel technology, which is addressed as WarmRAC, is both economically and environmentally sound [26]. Although these mixes have been used since the 1960s in the US [27], their technology is still evolving. Research has shown a satisfactory functionality of these pavements for improving aging, oxidation, fatigue, skid, reflection cracking, and rutting resistance as well as mitigating noise generation [26], improved durability and lower maintenance costs [28, 29]. To enhance various aspects of warm mix pavements, the research findings can be categorized into design, construction, performance, and recycling [26] as shown in **Table 6** (data from [26]).

6.1.2 Potentials for improvement in warm mix asphalt technology

Warm mix asphalt technology has been evolving since they were introduced [30–33]. The prerequirement of reducing the energy needed for processing and compacting the mat is reducing the shear resistance of the mixture. Research shows that the primary approach in extending asphalt pavement's durability is the prevention of fatigue in them resulting from repeated loading and unloading by the traffic [33, 34]. The findings [11] indicate that warm mix asphalts have quite satisfactory performance in resisting cracking caused by fatigue [8].

1. Further reduction of required temperature for processing. As it was explained above, the reduction of the temperature in proceeding warm mix asphalts is proportional to energy consumption reduction, fewer fuel costs, lowered emission, and mitigated pollutions. Therefore, research must invest in further reduction of the processing temperature of these mixes. This reduction must maintain the expected mechanical properties of mixes. Optimizing the trade-off between lower temperature and the mechanical performance required further research.

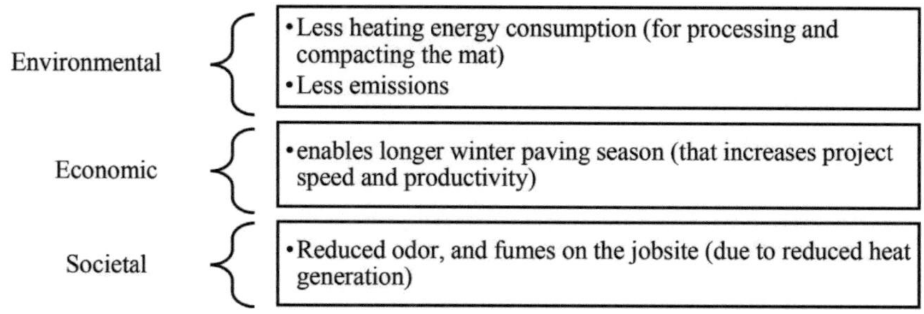

Figure 7.
Sustainability advantages of warm mix asphalt.

Main category	Subcategories	Potentials for improvement
Design	Material selection	Responsiveness to climate Responsiveness to high loads of traffic
	Mix gradation	Gap and open gradation Using in dense graded mixtures
	Optimum bitumen content selection	The relationship between bitumen content and durability, permeability, water susceptibility, and compaction
	Laboratory performance evaluation	Enhancing lab methods for conditioning/curing of test samples, workability, etc.
Construction	Temperature issues	Solving temperature issues during construction (through reducing the viscosity of binder)
	Production rate	Increasing the production rate to increase the construction speed
	Compaction	Achieving the target compaction density of WMA
Performance	Comparing different types of WarmRAC	Warm rubberized asphalt binder, Warm mix rubberized asphalt concrete, etc.
Recycling	Recycling methods for WarmRAC	The differences between recycling WarmRAC and conventional mixes Increasing the reusability
Environmental analysis	Potential environmental effects of RAC pavements	Air quality and occupational exposure Water quality

Table 6.
Research approaches in enhancing the warm mix asphalt technology.

2. Incorporating RAP in warm mix asphalt. Both warm mix asphalts and RAP are the results of sustainable approaches in asphalt production. The combination of both approaches is expected to maximize the sustainability of asphalt production. These innovative warm mixes are recycled asphalt with bitumen emulsion. Applying them is expected to contribute to both the economy and the environment. The lowered temperature allows using a larger amount of materials for recycling. This can reduce material use and increase energy savings and reduce costs [35]. A variety of lab tests have been conducted to evaluate the mechanical properties of these mixes including fatigue, bending, rutting, Marshall immersion, aging, and freeze-thaw-cycle splitting. The findings show that application of RAP can negatively impact the moisture and low-temperature cracking resistance, rutting resistance, aging, tensile strength ratio (TSR), and fatigue resistance [36]. On the other hand, there are findings that indicated an improved performance of mixes that use RAP [37]. Further research is required to resolve the possible poor performance of these mixes considering their significant contributions to sustainability.

6.2 Reducing the transfer of pollutants through runoffs by using permeable asphalt pavements

One of the primary negative impacts of the expansion of asphalt pavements is modifying the natural conditions of runoff paths by extending impervious surfaces that cannot effectively capture runoffs [38]. Stormwater runoffs contribute to various environmental issues including pollution of water bodies, reduction of groundwater recharge, and increasing the rivers' water temperature [39]. According to USEPA, urban runoffs are the leading source of water quality impairment to surveyed estuaries and the third-largest source of impairments to surveyed lakes [40]. Considering these significant detrimental impacts on the environment, the asphalt industry has an indispensable role in developing asphalt design and production modifications that contribute to stormwater runoff control. A primary approach in doing so is increasing the runoff capturing capacity of pavements (**Figure 8**).

6.2.1 The sustainability advantages of using permeable asphalt pavements

Permeable, pervious, or porous asphalt pavements have been identified as one of the dominant solutions and best management practices (BMPs) in controlling urban runoffs. They assist the transfer of runoffs to drain through the layers of the pavement into the ground below. The technology of permeable asphalt is still evolving, and significant ongoing research is currently followed to enhance the functionality of these pavements in different aspects such as durability, runoff capturing capacity, and ease of maintenance. Some of the sustainability advantages of using permeable pavements are shown in **Figure 9** [41].

As it can be seen (from **Figure 9**), the advantages of pervious pavements are significant, and a vigorous attempt must be made to encourage the asphalt industry to replace impervious pavements with permeable pavements wherever they can be applied. While this transition may be costly, the savings that can be made from eliminating the need to make more expensive stormwater runoff control devices such as retention ponds and swales, specifically in urban areas where land is very expensive, compensate the additional costs [42].

Additionally, the application of permeable asphalt pavements is more critical in urban areas for two reasons. One is the rapid growth of cities that is accompanied by replacing the soil and natural runoff paths with concrete and asphalt pavements [43]. According to the United Nations' report, in 2010, more than half of the earth's population lived in cities [44]. According to the research, between 60 and 100% of

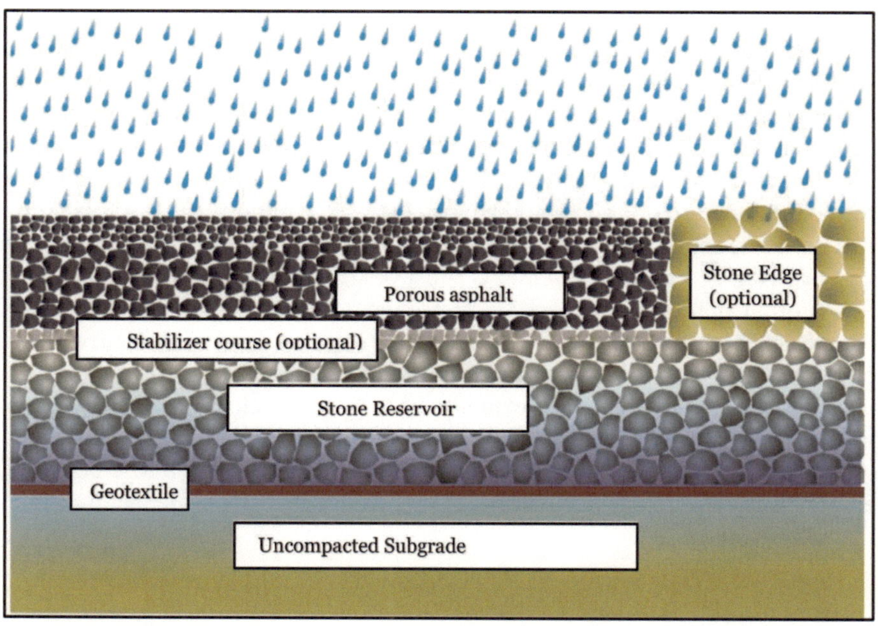

Figure 8.
Typical porous asphalt pavement with stone reservoir cross section (courtesy of Federal Highway Administration).

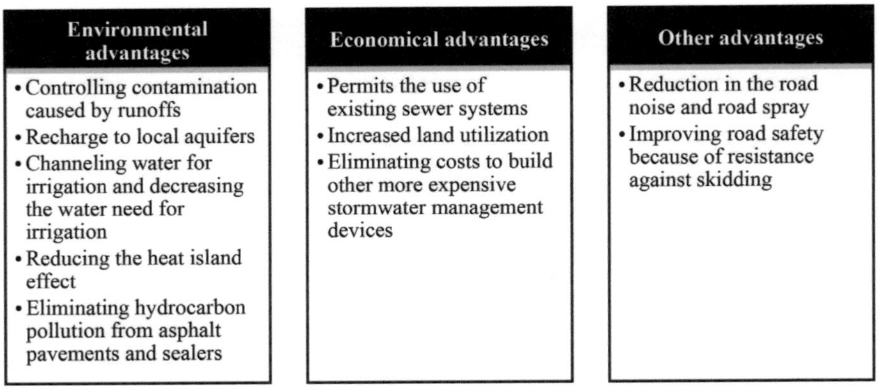

Figure 9.
The advantages of using permeable pavements for the triple sustainability bottom lines.

Figure 10.
Potentials for pervious asphalt pavements' mechanical property advancement.

the total urban areas have impermeable surfaces [45]. This is exacerbated by the multiple negative impacts of construction projects with long and often extended project durations [46, 47]. The second reason is the high presence of chemical pollutants and particles in urban runoffs [48]. This can be attributed to housing, vehicles, and industry activities in cities [49–51].

6.2.2 Research status of permeable asphalt pavements and filling the knowledge gap

Although there are adequate reasons for increasing the application of permeable pavements, there are certain barriers that limit their application. The research in the asphalt industry should focus on minimizing these barriers. The predominant barrier with this regard is the reduced resistance of permeable pavements against high loads of traffic. Considering that parking lots are one of the sensitive urban areas to collect polluted runoffs, the low compressive strength of permeable pavements limit their application for the parking lots where heavy equipment tools are parked. Also, the resistance of these pavements must be improved to withstand high-speed traffic. Another potential for mechanical property improvement is the permeable pavement's resistance against freeze-thaw as these pavements can capture and store water particularly when they are poorly constructed, and the infiltration process in them takes longer than it should normally take. Also, the clogging of these pavements over time and as a result of exposure to fine particles reduced their functionality [41]. **Figure 10** shows four potentials for improvement in the permeable pavements that require further research.

7. Summary and discussion

This chapter discusses the instrumental role of expanding asphalt pavements in nations' growth and development and the enormous and ever-rising demand to asphalt. It was stated that although asphalt production, in comparison, with other types of pavements is not considered an unsustainable industry, due to massive volumes of its production, it can have significant impacts on the sustainability triple bottom line and, more particularly, on the environment. The main sustainability in this regard was consumption of virgin materials, energy, and water as well as emission of pollutants. A key solution for pushing asphalt industry was suggested to be a comprehensive life-cycle analysis to identify all the possible potentials to implement sustainable practices in the four phases of material procurement, processing, transportation, and end of life point. Some policy recommendations were provided in each category. This chapter also discusses the opportunities for further advancement of the current sustainable approaches in the asphalt industry. The two main approaches were warm mix asphalts and pervious asphalt pavements. It was explained that although these technologies have significant sustainability advantages, there are some drawbacks in their application that need to be resolved through a collaborative effort of the industry and research. Some of the potentials of mechanical improvements of the two technologies were mentioned. It is essential that the individuals involved in the asphalt industry understand the essentiality of applying sustainable practices, gain enough knowledge and expertise of how to implement them, and then pursue the application of sustainable practices throughout the production process. Researchers, on the other hand, are responsible to target mitigating the existing environmental risks of asphalt pavements.

Author details

Mohammadsoroush Tafazzoli
Washington State University, Pullman, WA, United States of America

*Address all correspondence to: tafazzoli@wus.edu

References

[1] Santero NJ, Horvath A. Global warming potential of pavements. Environmental Research Letters. 2009; **4**(3):034011

[2] Asphalt Pavement Alliance. Asphalt is the Sustainable Material for Constructing Pavements. 2019. Available from: http://www.asphaltroads.org/why-asphalt/environment/ [Accessed: April 14, 2019]

[3] Mazumder M, Sriraman V, Kim HH, Lee SJ. Quantifying the environmental burdens of the hot mix asphalt (HMA) pavements and the production of warm mix asphalt (WMA). International Journal of Pavement Research and Technology. 2016;**9**(3):190-201

[4] Rebitzer G, Ekvall T, Frischknecht R, Hunkler D, Norris T, Rydberg W, et al. Life cycle assessment, part 1: Framework, goal and scope definition, inventory analysis, and applications. Environment International. 2004;**30**: 701-720

[5] Mukherjee A. Life cycle assessment of asphalt mixtures in support of an environmental product declaration. In: National Asphalt Pavement Institution. Maryland: Lanham; 2016

[6] Zapata P, Gambatese JA. Energy consumption of asphalt and reinforced concrete pavement materials and construction. Journal of Infrastructure Systems. 2005;**11**(1):9-20

[7] Butt AA, Mirzadeh I, Toller S, Birgisson B. Life cycle assessment framework for asphalt pavements: Methods to calculate and allocate energy of binder and additives. International Journal of Pavement Engineering. 2014; **15**(4):290-302

[8] Dinis-Almeida M, Afonso ML. Warm mix recycled asphalt—A sustainable solution. Journal of Cleaner Production. 2015;**107**:310-316

[9] D'Angelo J, Harm E, Bartoszek J, Baumgardner G, Corrigan M, Cowsert J, et al. Warm-Mix Asphalt: Europe Practice. Federal Highway Administration. U.S. DoT, AASHTO, NCHRP, Report No. FHWA PL-08-007; Alexandria; 2008

[10] Capitão SD, Picado-Santos LG, Martinho F. Pavement engineering materials: Review on the use of warm-mix asphalt. Construction and Building Materials. 2012;**36**: 1016-1024

[11] Carvalho T, Barreno IP. Ligantes betuminosos temperados. In: VII. Congresso Rodovi-Ario Portugues e Novos Desafios Para a Atividade Rodovi-Aria; 10-12 April 2013; Lisboa, Portugal (in Portuguese). 2013

[12] Miró R, Valdés G, Martínez A, Segura P, Rodríguez C. Evaluation of high modulus mixture behaviour with high reclaimed asphalt pavement (RAP) percentages for sustainable road construction. Construction and Building Materials. 2011;**25**(10):3854-3862

[13] Croteau JM. Performance and cost benefits of asphalt pavement recycling. In: Semi-Annual Meeting: Asphalt Recycling and Reclaiming Association (ARRA). 2005

[14] Chesner WH, Collins RJ, MacKay MH, Emery J. User Guidelines for Waste and by-Product Materials in Pavement Construction, No. FHWA-RD-97-148, Guideline Manual, Rept No. 480017. Recycled Materials Resource Center; 2002

[15] EAPA. Asphalt in Figures 2013. Brussels, Belgium: European Asphalt Pavement Association; 2014

[16] Silva HM, Oliveira JR, Jesus CM. Are totally recycled hot mix asphalts a sustainable alternative for road paving? Resources, Conservation and Recycling. 2012;**60**:38-48

[17] Lake R. Bottled Water Statistics: 23 Outrageous Facts. 2015. Available from: https://www.creditdonkey.com/bottled-water-statistics.html [Accessed: April 14, 2019]

[18] Ahmadinia E, Zargar M, Karim MR, Abdelaziz M, Ahmadinia E. Performance evaluation of utilization of waste polyethylene terephthalate (PET) in stone mastic asphalt. Construction and Building Materials. 2012;**36**:984-989

[19] Rahman WMNWA, Wahab AFA. Green pavement using recycled polyethylene terephthalate (PET) as partial fine aggregates replacement in modified asphalt. Procedia Engineering. 2013;**53**:124-128

[20] EAPA. The Use ofWarm Mix Asphalt. Brussels, Belgium: European Asphalt Pavement Association; 2010

[21] Oliveira JRM, Silva HMRD, Abreu LPF, Fernades SRM. Use of a warm mix asphalt additive to reduce the production temperatures and to improve the performance of asphalt rubber mixtures. Journal of Cleaner Production. 2013;**41**:15-22

[22] Hassan M. Evaluation of the environmental and economic impacts of warm-mix asphalt using life-cycle assessment. International Journal of Construction Education and Research. 2010;**6**(3):238-250

[23] Johnston A, Da Silva M, Soeymani H, Yeung CK. An early evaluation of the initial Canadian experience with warm asphalt mix. In: Annual General Conference of the Canadian Society for Civil Engineering; Calgary, Alberta, Canada. 2006

[24] Newcomb D. Warm mix: The wave of the future. In: HMAT: Hot Mix Asphalt Technology. Vol. 10. 2005. p. 4

[25] Hurley GC, Prowell BD. Evaluation of potential processes for use in warm mix asphalt. Journal of the Association of Asphalt Paving Technologists. 2006; **75**:41-90

[26] Wang H, Liu X, Apostolidis P, Scarpas T. Review of warm mix rubberized asphalt concrete: Towards a sustainable paving technology. Journal of Cleaner Production. 2018;**177**: 302-314

[27] Epps LA. Thermal Behavior of Crumb-Rubber Modified Asphalt Concrete Mixtures. Berkeley, California, USA: Civil Engineering, Univeristy of California; 1997

[28] Lo Presti D. Recycled tyre rubber modified bitumens for road asphalt mixtures: A literature review. Construction and Building Materials. 2013;**49**:863-881

[29] Shu X, Huang BS. Recycling of waste tire rubber in asphalt and Portland cement concrete: An overview. Construction and Building Materials. 2014;**67**:217-224

[30] Kheradmand B, Muniandy R, Hua LT, Yunus R, Solouki A. An overview of the emerging warm mix asphalt technology. International Journal of Pavement Engineering. 2014;**15**(1): 79-94

[31] Prowell B, Hurley G, Frank B. Warm-Mix Asphalt: Best Practices. 2nd ed. National Asphalt Pavement Association: Quality Improvement Publication; 2008. p. 125

[32] Rubio MC, Martínez G, Baena L, Moreno F. Warm mix asphalt: An overview. Journal of Cleaner Production. 2012;**24**:76-84

[33] Zaumanis M. Warm mix asphalt investigation [Master of Science thesis]. Kgs. Lyngby: Technical University of Denmark in cooperation with the Danish Road Institute, Department of Civil Engineering; 2010

[34] Teixeira A. Caracterização da resistência a fadiga de misturas betuminosas em equipamento servopneum atico [MSc thesis] (in Portuguese). Porto, Portugal: Faculdade de Engenharia da Universidade do Porto; 2000

[35] Dinis-Almeida M, Castro-Gomes J, de Lurdes Antunes M. Mix design considerations for warm mix recycled asphalt with bitumen emulsion. Construction and Building Materials. 2012;**28**(1):687-693

[36] Guo N, You Z, Zhao Y, Tan Y, Diab A. Laboratory performance of warm mix asphalt containing recycled asphalt mixtures. Construction and Building Materials. 2014;**64**:141-149

[37] Sengoz B, Oylumluoglu J. Utilization of recycled asphalt concrete with different warm mix asphalt additives prepared with different penetration grades bitumen. Construction and Building Materials. 2013;**45**:173-183

[38] Tafazzoli M. Investigating the Impacts of Green Roofs' Vegetation Properties on their Function in Controlling Urban Runoffs. International Low Impact Development Conference2018. pp. 176-183

[39] Driscoll E, Shelley PE, Strecker EW. Pollutant Loadings and Impacts from Highway Stormwater Runoff, Vol. I–IV, FHWA/RD-88-006-9. Oakland, Calif: Federal Highway Administration, Woodward-Clyde Consultants; 1990

[40] USEPA. Protecting Water Quality from Urban Runoff. EPA 841- F-03-003. Washington, DC: USEPA; 2003

[41] Tafazzoli M. Enhancing the Functionality of Pervious Concrete Pavements through Design and Maintenance. International Low Impact Development Conference2018. pp. 184-192

[42] Tennis PD, Leming ML, Akers DJ. Pervious Concrete Pavements. No. PCA, Serial No. 2828. Skokie, IL: Portland Cement Association; 2004

[43] Bates B, Kundzewicz Z, Wu S. Climate Change and Water. Intergovernmental Panel on Climate Change Secretariat. Geneva: IPCC Secretariat; 2008. p. 210

[44] United Nations. Population Distribution, Urbanization, Internal Migration and Development: An International Perspective. Department of Economic and Social Affairs Population Division; 2011

[45] Lu D, Weng Q. Use of impervious surface in urban land-use classification. Remote Sensing of Environment. 2006; **102**(1–2):146-160

[46] Tafazzoli M. Strategizing sustainable infrastructure asset management in developing countries. In: International Conference on Sustainable Infrastructure. 2017. pp. 375-387

[47] Tafazzoli M, Shrestha PP. Investigating causes of delay in US construction projects. In: 53rd ASC Annual International Conference Proceedings, Associated Schools of Construction. 2017

[48] Jartun M, Ottesen RT, Steinnes E, Volden T. Runoff of particle bound pollutants from urban impervious surfaces studied by analysis of sediments from stormwater traps. Science of the Total Environment. 2008; **396**(2–3):147-163

[49] Kelly J, Thornton I, Simpson PR. Urban geochemistry: A study of the influence of anthropogenic activity on the heavy metal content of soils in traditionally industrial and non-industrial areas of Britain. Applied Geochemistry. 1996;**11**(1–2):363-370

[50] Mielke HW, Gonzales CR, Smith MK, Mielke PW. The urban environment and children's health: Soils as an integrator of lead, zinc, and cadmium in New Orleans, Louisiana, USA. Environmental Research. 1999; **81**(2):117-129

[51] Mielke HW, Wang G, Gonzales CR, Powell ET, Le B, Quach VN. PAHs and metals in the soils of inner-city and suburban New Orleans, Louisiana, USA. Environmental Toxicology and Pharmacology. 2004;(3):243-247

Section 3

Asphalt Aging and Rejuvenation

Chapter 5

The Evaluation of Asphalt Mixture Mastic as an Aging Indicator

Carlos Alfonso Cuadro Causil, Wilmar Darío Fernández-Gómez, Jorge Iván Osorio Esquivel and Fredy Alberto Reyes Lizcano

Abstract

This chapter studies two mastic rheological properties after aging processes. The former is mastic produced in the laboratory, and the latter is the one extracted from asphalt pavements in service. A Colombian asphalt cement of two penetration grades was used to mix laboratory mastics, and two fillers were used also. Field mastics were extracted from in-service asphalt pavements constructed with the same asphalt mixture. Laboratory mastics were submitted to varying accelerated aging treatments in the laboratory, rolling thin-film oven (RTFO) test and an ultraviolet (UV) radiation chamber, to short- and long-term aging, respectively. Rheological measurements with a dynamic shear rheometer (DSR) were developed and calculated the aging master curves based on the rheological models. The results show the progression of hardening in the laboratory mastics as well as field mastics. In summary, mastic analysis is going to be a good indicator of the asphalt mixture aging.

Keywords: aging, rheology, mastic, SHRP aging, UV radiation

1. Introduction

Highway pavement asphalt is affected by the presence of oxygen, UV radiation, and temperature changes. Taken together, these three aging aspects appear throughout the production of the asphalt mixture, construction of the road, and service life. Asphalt's aging variables could be classified as either intrinsic or extrinsic. While the former refers to the volatilization of lightweight fractions, oxidation (oxygen absorption and diffusion), thixotropy, syneresis, and degradation of the polymeric structure (in the case of modified asphalts), the latter refers to the mixture's air void content, binder's film thickness, transit and thermal wear, as well as the aggregate type and filler proportions (filler minerals or fine materials). An undoubtedly complex phenomenon, asphalt aging, has consequences ranging from reduced durability to changes in the asphalt's physicochemical properties and chemical composition [1].

The production, storage, transport, and laying of asphalt mixtures cause compositional changes, i.e., changes in the mixture's physicochemical properties; these changes are also found in the asphalt mixture employed as road pavement [2–4]. However, the majority of design models for asphalt mixtures only focus on one of the "unaged" asphalt's physical properties. Moreover, within the variables and considerations, these models utilizing temporal changes in asphalt properties are not taken into account [5].

Not only it is faced with the difficulty concomitant to ensuring satisfactory yield from mixtures after the aggregates are added to the asphalt, but it must account for the incredible demand placed upon this mixture by long-term transportation. It is incumbent upon us, then, to employ laboratory techniques in a timely, cost-efficient manner to study the aging behavior of asphalt mixtures. Such techniques would ideally require the minimal amount of material for analysis. Consequently, mastics before the start of their service life were examined. Considering that the present data in the case of asphalt aging via mastic analysis is relatively scarce (in comparison to that of asphalt mixtures in general), this study contributes new data: the effect of mastics in the long-term aging of asphalt mixtures.

The physicochemical properties of mastic directly and significantly influence the performance of asphalt mixtures based on the type of filler mineral used and the relative proportions of the mixture itself—variables which could factor heavily in the mixture's resistance to aging. This research aims to fill the gap in knowledge with regard to the effect of filler minerals on the long-term aging of asphalt binders [5].

The efficacy of basing predictions for asphalt mixture behavior on mastics by way of a rheological analysis of the latter using laboratory-aged asphalt mixture samples developed with Colombian ACs 60-70 and 80-100. This project, then, addresses the need for the development of an asphalt aging index (AI) based on the rheological analysis of mastics. To do so, both field and laboratory samples were analyzed. In sum, the benefit of mastic analysis lies in that it allows to construct master curves and predict the asphalt mixture aging.

The specific objectives of this project are as follows:

Employ rheological analyses on mastics produced in the laboratory.

Construct master aging curves with the data obtained from asphalt mastics in the field.

Propose a linear viscoelastic model of asphalt mixture behavior that has a quadripartite scientific foundation: the linear viscoelastic model, the effect of aging, the aging master curves, and the aging shift factor.

1.1 Mastic

In general, hot-mix asphalt (HMA) is composed of aggregate minerals, asphalt binder, and air voids. Insofar as the mastic is concerned, it is composed of binder and filler minerals (mineral aggregate put through a Sieve #200 or less than 75 μm). The fillers' mineral components play a crucial role in both the mastics' and the HMA's properties as they can diminish the aging effects by virtue of increased viscosity [6–8]. Therefore, a thorough understanding of the fillers in both the mastic and the HMA proves to be invaluable for developing high-performance, well-designed asphalt [9–12].

Various studies on the rheological properties of asphalt mastic and other types of filler highlight the diversity of mineral fillers used, in addition to identifying the reinforcement they provide when added to asphalt cement. The consensus reached is that these fillers influence the physicochemical interactions within the mixture itself; in other words, filler is more than some inert substance used to fill holes between thick aggregate particles in the asphalt mixture. In fact, it is an active material whose activity occurs in the interface between the applied load and the asphalt binder [13]. Although Lesueur and Little [14], Hopman et al. [15], and Buttlar [8] have documented the aforementioned interactions, the role of fillers in asphalt mixtures is sufficiently complex so as to require further study.

1.2 Viscoelasticity

In a viscoelastic material, the relationship between stress and deformation depends on the weight and speed of the load applied. When materials of this type are faced with a load, they present a characteristic relaxation time that corresponds to the amount of time each material takes to restructure itself and achieve equilibrium. Due to the fact that each viscoelastic material has its own relaxation time dependent upon temperature and load application, a temperature increase translates into faster structural readjustment (i.e., temperature reduces relaxation time). Therefore, we can induce any material to reach the same relaxation time with either a quick high-temperature or a slow low-temperature treatment.

Myriad asphalt studies have observed its viscoelastic nature and concluded that fundamental rheological methods should be employed to obtain proper material characterization. Unfortunately, few have had the necessary resources and sufficient background to study asphalt films (binders) by way of rigorous rheological methods [16]. The characterization of the viscoelastic properties of asphalts uses a rheometer with cone and plate geometry, which measures the complex moduli of aged and unaged asphalts over a wide array of temperatures and frequencies. As a result, these studies hypothesize that oxidation changes the temperature dependence of the material and that these changes tend to increase as temperature increases [17]. Dickinson and Witt [18] later confirmed these findings.

1.3 Aging

Aging is usually quantified in the form of an Aging Index (AI), whether directly or indirectly. These aging indices have been especially popular in regard to the quantification of long-term aging (service life). A single AI point portrays the relation between aged and unaged asphalt viscosities, as well as describes increased hardening when the asphalt's response is essentially viscous. The problem here is that this single point does not offer an exact reflection of the changes in rigidity at low temperatures when the phase angle provides a significant portion of the response [19]. In 1981, Vallerga et al. used UV radiation and infrared lights to age asphalt binders. UV radiation turned out to be an extremely effective method—in empirical terms—of changing the asphalt's softening point, ductility, and penetration after oxidation had occurred; similar results were obtained by Fernández et al. [3].

1.4 Time-temperature superposition and master aging curves

As it has been mentioned, the behavior of viscoelastic material depends on temperature and load frequency: a viscoelastic material subject to high-frequency loads at low temperatures behaves the same as one subject to low-frequency loads at high temperatures. A change in temperature modifies the distribution of relaxation times such that all times corresponding to a distribution obtained at a temperature (*Td*) are related to the corresponding times of another distribution obtained at a different temperature (*T*). Thus, relation between the two relaxation times corresponds to the respective temperatures.

The dependence a_t on temperature is expressed by the Williams-Landel-Ferry (WLF) equation:

$$Log a_T = \frac{-C_1(T - Td)}{C_2 + T - Td} \tag{1}$$

where a(T) is the aging shift factor, T is the aging time for the present study, T_d is the reference aging time for the present study, and C_1 and C_2 are the empirical constants. In light of the correlation between aging effects and temperature, it developed the master aging curves similar to the master temperature curves by shifting the experimental data horizontally. Given that the rheological properties include the complex moduli and the phase angles, two master curves were needed to adequately describe the material's behavior [5].

Analysis of the shift factor in SHRP aged and unaged asphalts carried out by the authors of this study shows that the constants for the WLF equation can essentially assume the same values: −19 for C1 and 90 for C2. These values concur with those previously obtained by other authors [20], where the amount of shift required for each temperature to form the master curve is especially important and is referred to as the shift factor a(T). A graph Log a(T) against temperature is normally prepared in conjunction with the master curve. Such a graph visually indicates how the viscoelastic properties of the materials change with temperature.

Applying the principle of time-temperature superposition (TTS) to the construction of master curves, in addition to determining the shift factor, proves to be a powerful investigative tool that clearly outlines the time and temperature dependence of asphalt cements [21]. The temperature dependence in asphalt cements, in terms of the variation of Log a(T), is a function of time, and it can be mathematically modeled using the WLF equation at high temperatures.

Aging time data shifts with respect to the logarithm of time until the curves converge to form a single smooth one. The resulting master aging curve moduli describe the dependence of the material on aging time. The amount of shift for each aging time required for the master curve describes the material's dependence on aging time; the logarithm of the aging shift factor is the shift undergone by the complex moduli in relation to a given reference aging time (to form a single curve). The reference aging time to construct the master aging curve can be arbitrarily chosen [5].

1.5 Linear viscoelastic model

Asphalt's viscoelastic nature has led to a variety of ways to interpret the behavior of natural asphalts and asphalt mixtures. The models proposed range from the empirical (also known as mathematical or phenomenological models), e.g., Jongepier and Kuilman [22], Dobson, [23], Dickinson and Witt [18], Christensen and Anderson [21], Stastna et al. [24], Marasteanu and Anderson [25], Zeng et al. [26], Elsefi et al., and Mohammad et al. [27], to the mechanical (based on rheology), e.g., Huet [28], Huet-Sayegh model [29], Olard and Di Benedetto [30], and Di Benedetto and Neifar [31]. The former set of models tries to predict the behavior of the viscoelastic material using master curves that combine complex modulus and phase angle behavior within a wide range of frequencies and temperatures. In contrast, the latter set of models takes into account the material's linear and nonlinear behavior. In general, these models are capable of predicting the linear viscoelastic properties by employing the TTS principle. For a detailed description of the different models, see Yusoff et al., [32].

2. Materials and methods

2.1 Materials

Two types of mastics were evaluated for this project: the first was developed in the laboratory, and the second was extracted from field pavements. The laboratory mastics were made with an asphalt cement (AC) of two penetration grades most commonly found in Colombia, AC-60-70 and AC-80-100, from the Ecopetrol Refinery located in Barrancabermeja, Colombia; they contain a fine material produced by the trituration of alluvial material and in proportions of 2, 10, and 20% of the mixture's total weight. From the combination of the two asphalt cements and the three proportions of fine material, six distinct samples were obtained (identified in **Tables 1–3**).

To determine the rheological characteristics of the two asphalts, a TA Instruments dynamic shear rheometer (DSR) AR 2000 EX was used. The evaluation yielded results according to the performance grade (ASTM D6373-07) for medium to high temperatures; **Tables 2** and **4** display those results. Asphalt characterization was led to classify both as PG 58 for high temperature and 16 for medium temperature; low temperature was not measured because pavements in Colombia are not exposed to temperatures below 0°C (**Table 5**).

The FTIR spectroscopy allowed us to map the effects of UV radiation on the asphalts studied. Due to the low sensibility concomitant to FTIR spectroscopy in macromolecules like asphalts, we used the oxidants H_2O_2 (30% v/v) and HNO3 (63% w/v) to help identify functional groups C═O or S═O. Due to the low sensibility concomitant to FTIR spectroscopy in macromolecules like asphalts, we used the oxidants H_2O_2 (30% v/v) and HNO_3 (63% w/v) that help us to identify on the functional groups the magnitude of strong oxidation C═O or S═O. The oxidants mixed with original asphalts AC-20 and AC-30 for 48 h were tested by infrared

Sample	Code
AC-60-70 + 2% filler mineral	1
AC-60-70 + 10% filler mineral	2
AC-60-70 + 20% filler mineral	3
AC-80-100 + 2% filler mineral	4
AC-80-100 + 10% filler mineral	5
AC-80-100 + 20% filler mineral	6

Table 1.
Laboratory samples.

Experiment	Method	Unit	AC-60-70
Original binder			
Penetration (25°C, 100 g, 5 s)	ASTM D-5	0.1 mm	63.4
Specific weight	INV. E-707	—	1.012
Penetration index	NLT 181/88	—	0
DSR viscosity (60°C)	ASTM D-4402	Pa s	145.6
Ductility (25°C, 5 cm/min)	ASTM D-113	cm	>105
Softening point	ASTM D-36-95	°C	52.4
Solubility trichloroethylene	ASTM D-2042	%	>99
Water content	ASTM D-95	%	<0.2
Flash point	ASTM D-92	°C	323
Residue RTFO test binder			
Mass loss	ASTM D-2872	%	0.3
Penetration (25°C, 100 g, 5 s)	ASTM D-5	0.1 mm	62

Table 2.
Physical characterization of AC-60-70.

Experiment	Method	Unit	AC-80-100
Original binder			
Penetration (25°C, 100 g, 5 s)	ASTM D-5	0.1 mm	83.2
Specific weight	INV E-707	—	1.007
Penetration index	NLT 181/88	—	0.3
DSR viscosity (60°C)	ASTM D-4402	Pa s	136
Ductility (25°C, 5 cm/min)	ASTM D-113	cm	>105
Softening point	ASTM D-36-95	°C	50.5
Solubility trichloroethylene	ASTM D-2042	%	>99
Water content	ASTM D-95	%	<0.2
Flash point	ASTM D-92	°C	358
Residue RTFO test binder			
Mass loss	ASTM D-2872	%	0.5
Penetration (25°C, 100 g, 5 s)	ASTM D-5	0.1 mm	61

Table 3.
Physical characterization of Colombian AC-80-100.

Temp.	Delta	G*	(G*/Sin δ)	(G*Sin δ)
°C	°	kPa	kPa	MPa
Original				
52	84.52	4.76	4.78	
58	86.18	1.90	**1.90**	
60	86.61	1.46	1.46	
64	87.40	0.82	0.82	
RTFO				
52	81.24	5.25	5.31	
58	82.10	3.98	4.02	
60	83.52	2.30	**2.32**	
64	85.38	1.09	1.09	
76	86.76	0.52	0.52	
PAV				
13	42.09	12.00		7.97
16	45.40	7.01		**4.84**
19	48.09	4.78		3.43
22	54.39	2.96		2.21

Table 4.
Partial performance grade for AC-20 (60-70).

spectroscopy and allowed us to identify changes in carbonyl and sulfoxide groups. The results shown in **Figure 1** exhibit significant increases in the 1650 cm^{-1} band, the domain of functional carbonyl groups, which indicates the oxidation of carbon atoms. We found minimal sulfoxide variations after oxidation, a situation

Temp.	Delta	G*	(G*/Sin δ)	(G*Sin δ)
°C	°	kPa	kPa	MPa
Original				
52	86.00	4.31	4.32	
58	87.22	1.81	1.81	
60	87.64	1.35	1.36	
64	88.09	0.91	0.91	
RTFO				
52	82.28	9.08	9.16	
58	84.32	3.90	3.91	
60	85.04	2.88	2.89	
64	85.98	1.72	1.72	
76				
PAV				
10	39.62	18.87		12.03
13	42.09	13.06		8.75
n16	45.40	6.83		4.87
19	48.09	4.84		3.60
22	54.39	2.58		2.10

Table 5.
Partial performance grade for AC-20 (80-100).

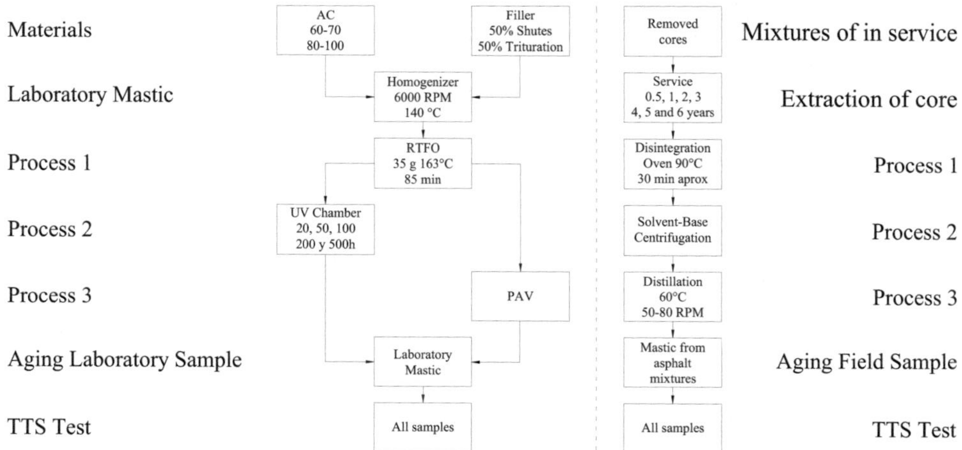

Figure 1.
FTIR spectra of original asphalts AC-30 and AC-20 (left) and AC-30 and AC-20 after 48 hours of oxidizing agent exposure (right) [33].

that leads us to propose that "unaged" asphalts exhibit S=O groups as a result of the previous aging brought about by extraction and refinement. **Figure 1** also allows us to compare the oxidation achieved by strong oxidants and that of the UV chamber [33].

In order to show an outline of the methods at this project, **Figure 2** plots an outline of this.

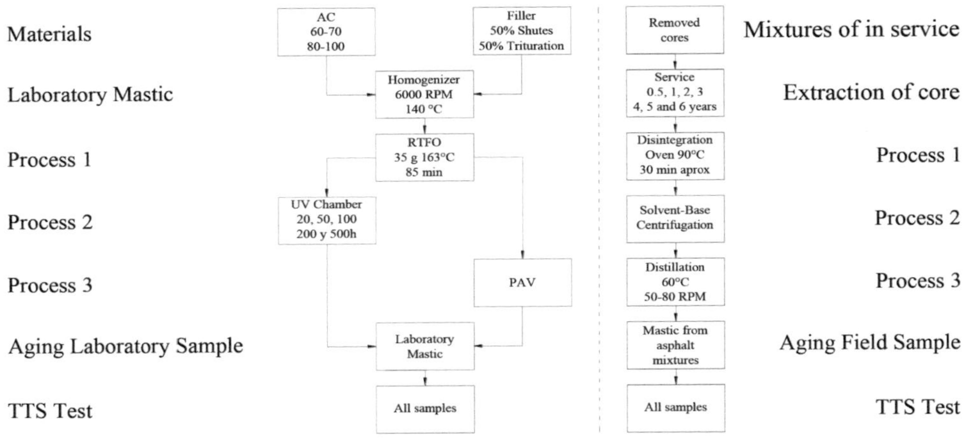

Figure 2.
Methodology outline [34].

Figure 3.
Filler aggregate (above) and chute aggregate (below).

The filler mineral is an aggregate from the Coello River in Tolima, Colombia. The filler was obtained by a mixture of 50-50 split between two sources: the first taken from the aggregate trituration plant and the second from the asphalt plant's chutes. **Figure 3** shows two different fillers. Before mixing the two materials, both are passed through a Sieve #200 (75 μm). After, the field mastic was extracted from flexible pavements used in Bogota's roads, with service life ages of 12, 24, 60, and 72 months. In order to obtain these field samples, a core of the asphalt binder

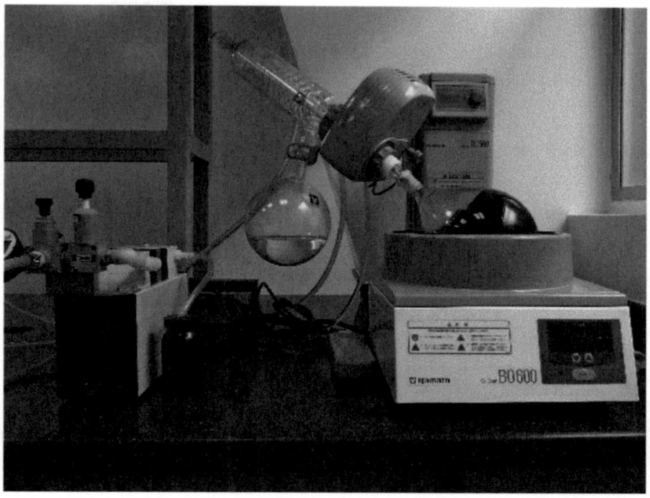

Figure 4.
Yamato BO600 rotary evaporator.

under study was extracted and brought to the laboratory. Thus, it can ensure that field mixture possesses the same gradation characteristics and asphalt content as that produced in the laboratory.

To fabricate laboratory mastic, the materials were heated at 140°C in a ventilation-free oven. The mixture is then produced with a HDD ULTRA TURRAX-T50 BASIC homogenizer for 5 minutes at a constant temperature of 140°C at 6000 rpm. The mixtures were stored in hermetic metal containers until tested. Once obtained, the fine asphalt mixture aggregate (the mastic) undergoes short- and long-term aging treatments. In the case of short-term treatment, a rolling thin-film oven (RTFO) Model CS 325-B manufactured by James Cox & Sons, Inc. was used for 85 minutes at 163°C. For the long-term aging treatment, the mastic was exposed to UV radiation in an aging chamber—ultraviolet ATL-360—that was built for this project following the standards established by ASTM D4799-08. Each exposure period consists of a 2-hour radiation cycle followed by a 2-hour condensation period, so that the mastic is subjected to continuous exposure periods of 20, 100, and 500 hours.

In order to recover the mastic from the asphalt mixtures of in-service pavements, cylindrical cores 10 cm in diameter and on average 10 cm thick were removed. Subsequently, disintegration, centrifugation, and solvent-based distillation of the core samples were carried out. The disintegration process consisted of heating the cores to 90°C for 40 minutes in a BLUE LINE OV 12A oven before gently separating the aggregates mechanically. The centrifugation was performed with a HOUGHTON MFG E2 centrifuge, using 800 g of asphalt mixture and 900 ml of solvents (70% toluene and 30% ethanol by volume). Centrifugation successfully achieved a mixture of solvents and mastics for each core sample. As far as distillation is concerned, the mixtures were placed in a Yamato BO600 rotary evaporator at a speed ranging from 50 to 80 rpm and with a temperature of 60°C. **Figure 4** shows the rotary evaporator. The entire distillation process takes approximately 30 minutes. Finally, the resultant mastic is left to "air out" for 3 days to facilitate the evaporation of the residual solvent and stored in hermetic metal containers until rheologically evaluated.

For both the laboratory and field samples obtained via the process outlined, the rheological evaluation of the materials was done with the DSR TA INSTRUMENTS AR-2000 EX's rheological testing. Tests were addressed on 25 mm geometry with a gap of 1000 μm at temperatures of 15, 25, 35, 45, 55, and 65°C. Taken together, this information and the TTS principle allow us to create master curves for behavior at a temperature of 25°C according to the aging shift factors produced by the CAM and WLF models.

2.2 UV radiation of the mixture

The samples were submitted to UV radiation in an aging chamber that has eight lamps. These lamps emit radiation with a wavelength of 340 nm in the range of UVA rays. **Figure 5** shows the aging chamber. Exposure times for the mastic samples analyzed varied between 0 and 500 hours to reflect conditions encountered in the field. In other words, if Bogota's average integrated solar radiation—measured in a wavelength of 340 nm—is 2.2 W/m^2 nm per day (IDEAM, 2005), we can extrapolate that figure to estimate an annual UVA exposure of 8.8 or 3155 kJ/m^2. Therefore, the aging chamber's radiation value of 1.55 W/m^2 nm per cycle (ASTM G 154-06) meant we had to put the mastic samples through 500 hours of UVA radiation to simulate a year of exposure in the field as pavement in Bogota (see **Table 6**). Within the chamber itself, we placed 2-mm-thick mastic samples on two 300-cm rectangular plates (10 cm wide and 3 mm deep).

The mastics were treated in the UV chamber's 4-hour cycles of radiation and condensation (as it has been previously stated, each cycle consists of 2 hours of radiation followed by 2 hours of condensation). The 2-hour condensation period exposed the samples to a temperature of 60°C at 99% humidity, since these conditions reflect Colombia's tropical climatic conditions. Moreover, they recreate conditions propitious for asphalt oxidation. Alternating between the two periods emulates what the materials face in the field. That is to say, after diurnal UV exposure, there comes a nocturnal "rest period" during which the

Figure 5.
Aging chamber.

UVA exposure period (hours)	Equivalency (months)
50	1.2
100	2.4
200	4.8
500	12

Table 6.
Exposure to UV radiation and months in the field equivalency.

reactions that determine the physical and chemical changes of the material occur [35]. The samples treated in the aging chamber were analyzed with the DSR to establish the changes exhibited by the asphalt—the difference between unaged and aged samples.

2.3 Time-temperature superposition test

The TTS test provides dynamic material testing over a number of frequencies at different temperatures. Owing to the fact that temperature corresponds to the linear viscoelastic region of the material, these curves are parallel, and the shift factor responds to the curve's horizontal shift toward a reference temperature at each given temperature (in this case, 25°C). This test essentially made use of the same

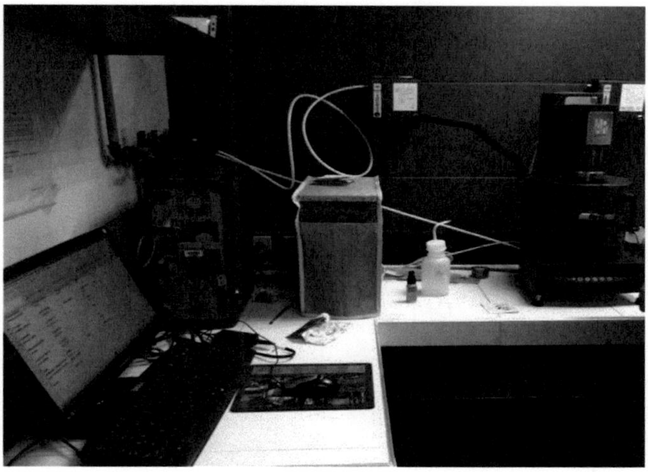

Figure 6.
DSR TA Instruments AR-2000 EX.

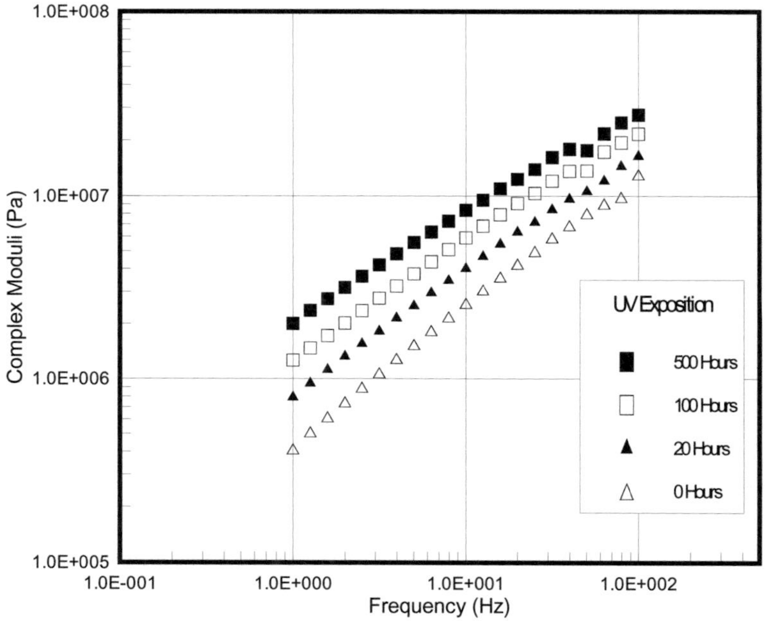

Figure 7.
Complex moduli (|G|) of sample 1 after various exposure periods.*

principle: instead of different temperatures, different aging periods were taken into account to obtain the aging shift factor for each sample.

The range of frequencies covered is from 1 to 100 Hz and that of temperatures is 25, 35, 45, 55, and 65°C. This investigation used a deformation percentage of 0.2% since mastic may present more hardening than the aged asphalt, and in order to ensure fidelity to the data over the linear viscoelastic region, those evaluations were performed with dynamic shear rheometer AR-2000 EX (**Figure 6**). From this information, master aging curves following the WLF model were created. The reference temperature was 25°C, and the curves' shift factors for four aging levels corresponding to 0, 20, 100, and 500 hours, respectively, were obtained. The graph below (**Figure 7**) displays the behavioral curves for the complex moduli of the AC-60-70's mastic with 2% of fillers at different stages of aging in the UV chamber. The same methodology was employed to construct the master curves for all six laboratory samples.

With the information gained from the TTS test, aging master curves were plotted using the modified CAM model, where empirical algebraic equations are employed for modified asphalts and asphalt mixtures under a dynamic load of low shear stress in a wide range of frequencies, temperatures, and deformations. The model is composed of four formulations for the master curves of |G*| and δ. Thus, equation simulates the behavior of the mastic both as a viscoelastic fluid and as a solid. The equation of the |G*| is rooted in a generalization of the CAM model and the universal model. Below, the reader will find the equations for the |G*|:

$$|G^*| = G_e + \frac{G_g - G_e}{\left[1 + \left(\frac{f_c}{f'}\right)^k\right]^{\frac{m_e}{k}}} \tag{2}$$

where $G_e = |G^*|(f \to 0)$ with $G_e = 0$ for mastic; $G_g = |G^*|(f \to \infty)$; f_c is a positional parameter with frequency dimensions; f' is the reduced frequency, a function of temperature and deformation; and k and m_e are dimensionless parameters.

3. Results and discussion

3.1 Aging master curves for the complex moduli in the laboratory

Because rheological properties encompass the complex moduli and phase angles, we were obligated to develop two master curves to properly describe the material's behavior. Here, it is necessary to emphasize that this research saw an increase in temperature that mirrored a decrease in aging time. Similar results have been reported [5]. The resulting shift is the aging master curve for the six samples presented in **Figure 8**.

As a result of our aging master curves, various shift factors were obtained; based on these factors, the empirical constants for the WLF model and the variables for the CAM model were calculated. The margin of error for the calculated empirical constants was less than 6%. The CAM model variables, the WLF empirical constants, and their correlations in all of the samples are summed up in **Table 7**.

Results were consistent with those of other studies insofar as the increase in aging time parallels that of the complex moduli as a factor in aging; for example, Huang [5] reports that increasing the level of aging time, the complex modulus increases. On the other hand, this author assumed E* (MPa) = 0 and m = 1 [5], in this investigation those variables were calculated, and our results are 0.0105 and 1.09, respectively.

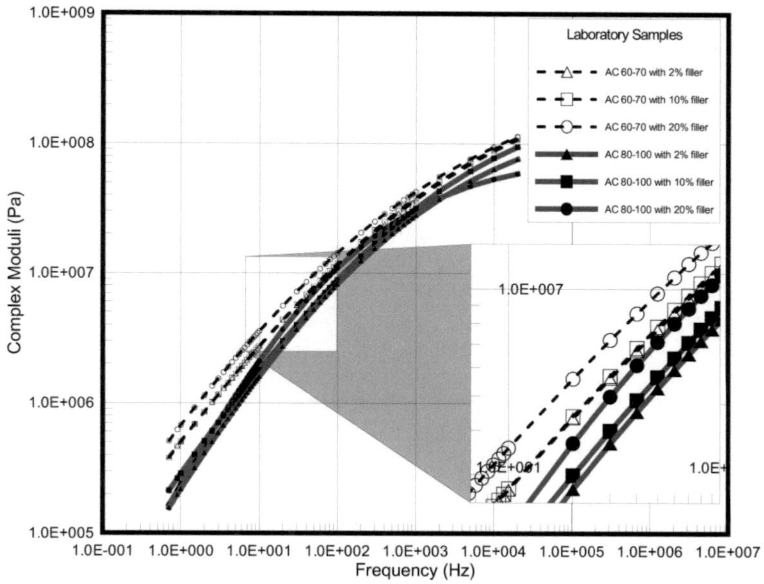

Figure 8.
Aging master curves for the six laboratory samples.

Sample	E*e (MPa)	E*g (MPa)	fc	k	me	R^2	C_1	C_2	R^2
1	0.011	424.03	220.32	0.22	1.04	0.995	0.85	53.30	0.987
2	0.011	424.03	180.45	0.22	1.03	0.995	0.75	17.06	0.988
3	0.013	427.03	157.05	0.23	1.02	0.987	0.86	59.61	0.940
4	0.002	201.68	191.99	0.28	1.14	0.995	0.78	126.24	0.943
5	0.010	267.45	154.30	0.26	1.17	0.996	0.80	119.71	0.993
6	0.001	82.34	81.35	0.39	1.21	0.994	0.71	17.98	0.999
Field	1.118	424.03	2.20	0.23	1.04	0.954	10	66.57	0.695

Table 7.
CAM and WLF constants.

Evidence of the aging dynamic is presented in **Figure 9**. This graphic charts the aging shift factor against aging time for each sample. The increase in shift factor as it relates to time could clearly identify. However, it must point out that the time presented corresponds to a long-term aging treatment. Therefore, the time scale can be interpreted in hours for laboratory aging or months in service life, as the equivalences in **Table 6** demonstrate. A logarithmic regression done with the R^2 coefficient was 0.75; this means that environmentally aged mastic can increase their complex modulus after 500 hours.

To validate the viscoelastic model, we refer readers to **Figure 10**, in which graphs of the complex moduli are measured against predictions. Strong correlation between the predicted and measured complex moduli was found (0.99).

The aging dynamic of the aging shift factor indicates aging. Hence, regressions can be used to predict the complex moduli that samples would present after 1000 hours of aging in the UV chamber. Nevertheless, the physical hardening present after aging depends on the aging shift factor and the slope of master curve in the linear viscoelastic region. From the CAM model, the m_e factor obtained corresponds to this slope. Readers can find the physical hardening (i.e., aging index) factors—especially as they relate to the aging shift factor for the mastics analyzed—in **Tables 8** and **9**.

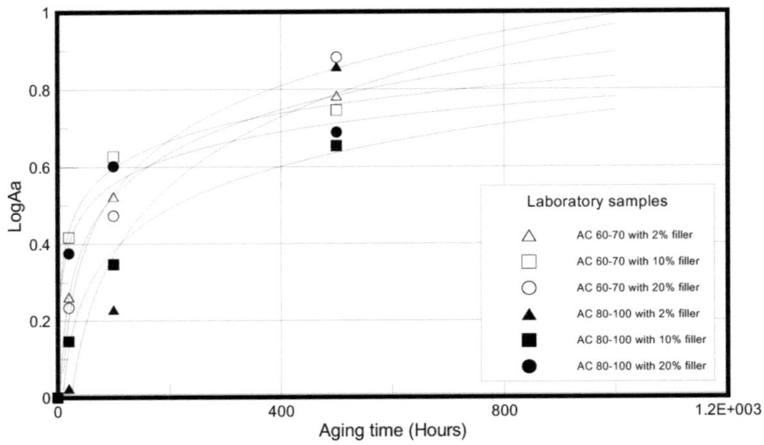

Figure 9.
Aging shift factors.

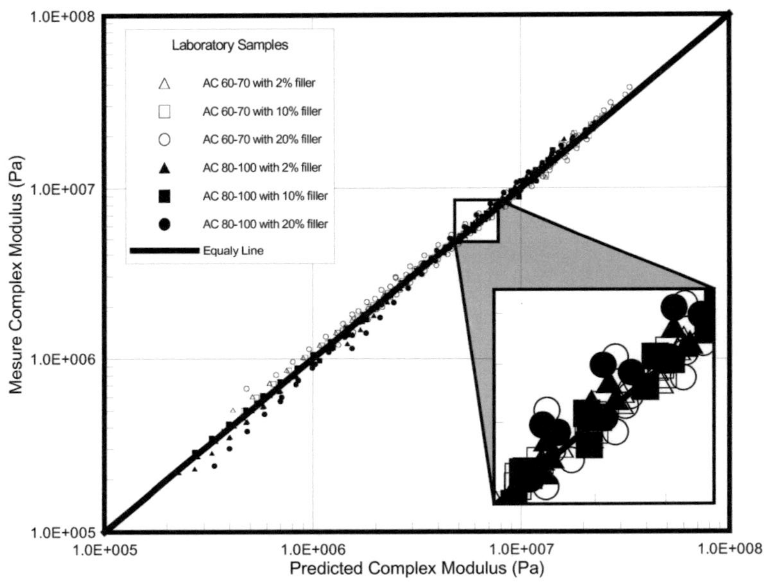

Figure 10.
Predicted versus measured complex moduli.

Age (h)	AC-60-70					
	ASF S-1	AI	ASF S-2	AI	ASF S-3	AI
me	1.04		1.03		1.02	
0	0	0	0	0	0	0
20	0.26	0.27	0.42	0.43	0.23	0.24
100	0.52	0.54	0.63	0.65	0.47	0.49
500	0.78	0.81	0.75	0.78	0.88	0.92

Source: Authors.

Table 8.
Mastic aging index AC-60-70.

AC-80-100						Field samples	
ASF S-4	AI	ASF S-5	AI	ASF S-6	AI	Age (years)	AI
1.14		1.17		1.21		1.04	
0	0	0	0	0	0	1	0.00
0.02	0.02	0.15	0.15	0.38	0.39	2	0.18
0.23	0.24	0.35	0.36	0.60	0.63	5	0.35
0.86	0.89	0.65	0.68	0.69	0.72	6	0.91

Table 9.
Mastic aging index AC-80-100 and field samples.

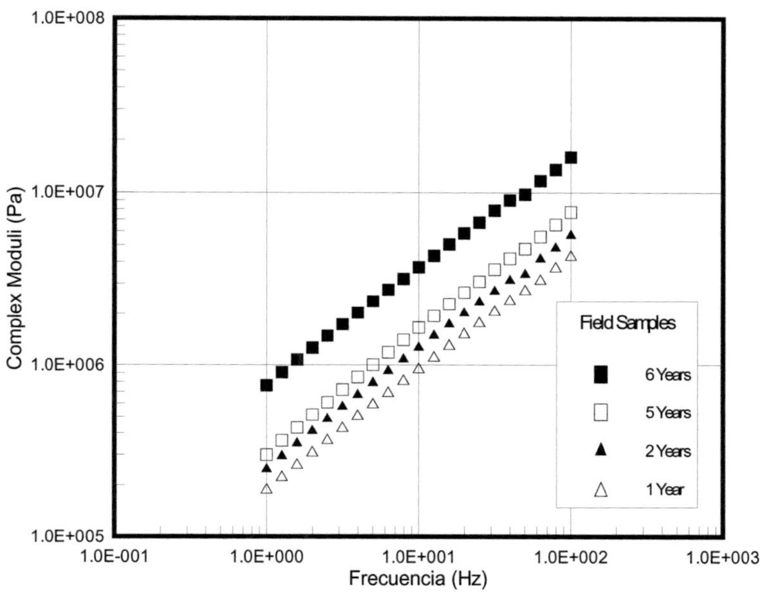

Figure 11.
Master curves for the complex moduli at 25°C.

3.2 Master curves for the complex moduli of field samples

The procedure used for the laboratory samples was also applied to the field samples. **Figure 11** charts the complex moduli of the field samples' master aging curves at different stages of service life.

If AI values superimpose on the data gleaned from the laboratory mastic, an acceptable correlation cannot be traced, perhaps due to the fact that the field mastic contains filler that solvent mixtures were unable to verify. Although behavior of 20% content fine's material is a good representative to field samples because correspond to average observed for those samples, future expectative research to correct this situation. Nonetheless, upon observing that the aging shift factor values approach 1 after 1 year of service life, we can state that the material exhibits its maximum hardening values, which do not increase substantially over time.

4. Conclusions

The rheological analysis of mastics allows us to predict aging in asphalt mixtures. The WLF equation for the temperature dependence of the complex moduli helps model mastic aging levels for Colombian asphalts 60-70 and 80-100, as it also does for the proportions of filler present in the mixture. The WLF model draws parallels between temperature and aging behavior: they have inverse proportions, with a decrease in temperature stimulating an increase in aging.

The prediction of the complex moduli after long-term aging for the laboratory samples proved to be acceptable. The proposed methodology permits researchers to obtain the Aging Index of asphalt mixtures by combining the aging shift factor and the slopes of the master curves (the latter established for multiple periods). To that end, the AI predicts mastic behavior—and, in turn, asphalt mixture behavior—during the first 1000 hours of UV chamber exposure or 2 years of service life. In Colombian mastics, hardening may increase up to 180% of the average, where the initial complex modulus serves as the starting value. This study also demonstrates that filler content reduces hardening in soft asphalts (AC-80-100) more than in hard asphalts (AC-60-70).

Colombian AC-60-70 exhibits worse performance in terms of aging in spite of the fact that the master curves of the complex moduli are similar to those of AC-80-100. This conclusion was based on observations that AC-60-70's AI is greater than that of AC-80-100 when filler content is 20%, as is often the case in the field. However, aging in AC-80-100 tends to progress more rapidly throughout service life, given that the slopes of its master curves are greater. These increased slopes translate into more susceptibility to cracking after the first year of service life.

The AI values presented by the field samples after 6 years of service life are parallel to those obtained from laboratory samples subjected to 500 hours of UV chamber treatment. Consequently, we believe aging chamber studies should continue to be carried out, so that we may collect data to determine whether the hardening process is continuous or whether environmental conditions cause softening-hardening cycles that went undetected by the procedures used.

Acknowledgements

The authors would like to thank the Material Testing and Civil Engineering laboratories at the Pontificia Universidad Javeriana; their gracious help was invaluable to the research conducted. In addition, thanks must go to Concrescol SA for providing the materials required to carry out our research. Lastly, the authors would like to thank Joe Wager for tidying up their English composition.

Author details

Carlos Alfonso Cuadro Causil[1*], Wilmar Darío Fernández-Gómez[2], Jorge Iván Osorio Esquivel[1] and Fredy Alberto Reyes Lizcano[3]

1 Pontificia Universidad Javeriana, Bogotá, Colombia

2 Center for Sustainable Pavements and Materials, Universidad Distrital Francisco José de Caldas, Bogotá, Colombia

3 Centro de Estudios de Carreteras, Transportes y Afines, Pontificia Universidad Javeriana, Bogotá, Colombia

*Address all correspondence to: carloscuadro@gmail.com

References

[1] Bianchetto H, Miró R, Pérez F. R esistencia al envejecimiento de las mezclas bituminosas en caliente: Beneficios y limitaciones de la incorporación de fílleres comerciales. Primera parte: estudios en base al método UCL; 2006

[2] Bell CA, Kliewer JE. Evaluating Aging of Asphalt Mixtures; 1995

[3] Fernández-Gómez W, Rondón H, Reyes F. A review of asphalt and asphalt mixture aging. Ingenieria e Investigación. 2013;**33**(1):5-12

[4] Gawel I, Baginska K. Effect of chemical nature on the susceptibility of asphalt to aging. Petroleum Science and Technology. 2004;**22**:1261-1271

[5] Huang S-C, Zeng M. Characterization of aging effect on rheological properties of asphalt-filler systems. International Journal of Pavement Engineering. 2007;**8**(3):213-223

[6] Bicerano J, Douglas JF, Brune DA. Model for the Viscosity of Particle Dispersions; 1999

[7] Butt AA, Jelagin D, Tasdemir Y, Birgisson B. The effect of wax modification on the performance of mastic asphalt. International Journal of Pavement Research and Technology. 2010;**3**(2):86-95

[8] Buttlar WG, Bozkurt D, Al-Khateeb GG, Waldhoff AS. Understanding asphalt mastic behavior through micromechanics. Transportation Research Record Journal of the Transportation Research Board. 1999;**1681**(1):157-169

[9] Anderson DA, Goetz WH. Mechanical Behavior and Reinforcement of Mineral Filler-Asphalt Mixtures: Technical Paper; 1973

[10] Cooley LA, Stroup-Gardinder M, Brown ER, Hanson DI, Fletcher MO. Characterization of asphalt-filler mortars with superpave binder tests. Journal of the Association of Asphalt Paving Technologists. 1998;**67**:42-66

[11] Harris BM, Stuart KD. Analysis of mineral fillers and mastics used in stone matrix asphalt (with discussion and closure). Journal of the Association of Asphalt Paving Technologists. 1995;**64**:54-95

[12] Huang B, Shu X, Chen X. Effects of mineral fillers on hot-mix asphalt laboratory-measured properties. International Journal of Pavement Engineering. 2007;**8**(1):1-9

[13] Kallas BF, Puzinauskas VP. A study of mineral fillers in asphalt paving mixtures. In: Proceedings, Association of Asphalt Paving Technologists. 1961

[14] Lesueur D, Little D. Effect of hydrated lime on rheology, fracture, and aging of bitumen. Transportation Research Record Journal of the Transportation Research Board. 1999;**1661**(1):93-105

[15] Hopman P, Vanelstraete A, Verhasselt A, Walter D. Effects of hydrated lime on the behaviour of mastics and on their construction ageing. In: Proceedings of the Durable and Safe Road Pavements, V International Conference; 11-12 May 1999; Held Kielce, Poland. Vol. 1. 1999

[16] Bahia HU, Anderson DA. The pressure aging vessel (PAV): A test to simulate reological changes due to field aging. In: Phys. Prop. Asph. Cem. Bind. 1995. pp. 67-88

[17] Sisko AW, Brunstrum LC. Relation of Asphalt Rheological Properties to

Pavement Durability, NCHRP Rep. No. 67; 1967

[18] Dickinson EJ, Witt HP. The dynamic shear modulus of paving asphalts as a function of frequency. Journal of Rheology. 1974;**18**:591

[19] Anderson DA et al. Binder Characterization and Evaluation. Vol. 3: Physical Characterization; 1994

[20] Anderson DA, Christensen DW, Bahia H. Physical properties of asphalt cement and the development of performance-related specifications. Journal of the Association of Asphalt Paving Technologists. 1991;**60**:437-475

[21] Christensen DW, Anderson DA. Interpretation of dynamic mechanical test data for paving grade asphalt cements (with discussion). Journal of the Association of Asphalt Paving Technologists. 1992;**61**:67-116

[22] Jongepier R, Kuilman B, Schmidt RJ, Puzinauskas VP, Rostler FS. Characteristics of the rheology of bitumens. In: Association of Asphalt Paving Technologists Proc. 1969

[23] Dobson GR. On the development of rational specifications for the rheological properties of bitumens. Journal of the Institute of Petroleum. 1972;**58**(559):24

[24] Stastna J, De Kee D, Powley MB. Complex viscosity as a generalized response function. Journal of Rheology. 1985;**29**:457

[25] Marasteanu MO, Anderson DA. Improved model for bitumen rheological characterization. In: Eurobitume Workshop on Performance Related Properties for Bituminous Binders. 1999

[26] Zeng M, Bahia H, Zhai H, Turner P. Rheological modeling of modified asphalt binders and mixtures. Asphalt Paving Technology. 2001;**70**:403-441

[27] Mohammad LN, Wu Z, Myers L, Cooper S, Abadie C. A practical look at the simple performance tests: Louisiana's experience (with discussion). Journal of the Association of Asphalt Paving Technologists. 2005;**74**:557-600

[28] Huet C. Étude par une methods d'impédance du comportement viscoélastique des matériaux hydrocarbons. Paris: Faculté des Science de Paris; 1963

[29] Sayegh G. Viscoelastic properties of bituminous mixtures. In: Intl Conf Struct Design Asphalt Pvmts. 1967

[30] Olard F, Di Benedetto H. The "DBN" model: A thermo-visco-elasto-plastic approach for pavement behavior modeling (with discussion). Journal of the Association of Asphalt Paving Technologists. 2005;**74**:791-828

[31] Kim YR. Modeling of Asphalt Concrete. ASCE Press United State of America; 2009

[32] Yusoff NIM, Shaw MT, Airey GD. Modelling the linear viscoelastic rheological properties of bituminous binders. Construction and Building Materials. 2011;**25**(5):2171-2189

[33] Fernandez-Gomez W, Rondón H, Daza C, Fredy R. The effects of environmental aging on Colombian asphalts. Fuel. 2014;**115**:321-328

[34] Cuadro Causil CA, Osorio Esquivel JI. Construcción de las curvas maestras del Mástic a partir del análisis reológico de muestras elaboradas en laboratorio y de muestras recuperadas de mezclas en servicio; 2012

[35] McGreer M. Weathering Testing Guidebook. Atlas Material Testing Technology; 2003

Section 4

Asphalt Pavement

Chapter 6

Microwave Processing of Mixtures of Asphalt and Şırnak Asphaltite Slime - Development of The Compression Ability, Shear and Tension Strength

Yıldırım İsmail Tosun

Abstract

The research studied the main consideration of the road construction industry as one of the current consumptions in road aggregate-asphaltite slime mixed asphalt production in Turkey. There was a great awareness of the source of opportunities that polymer-modified mixtures had for road asphalt production (PMA). The shear strength might be developed by microwave processing on compression or during compression. This study searched for the mechanical properties of many mixtures of asphaltite slime and co-polymers by microwave treatment and compression. This study presented the model processing method for the best road cover and showed an overview on the characteristics of polymer-asphalt binders, used in the paving industry. The main fact emphasized was shear resistivity and tension strength development by asphaltite slime made on the regressed relationships of strength and advanced properties of the binder. The relationships between the mixture components of the binder and compression performance of the mixture were also determined.

Keywords: microwave radiation, pyrolysis, Şırnak asphaltite, semi-molten salt, slurries, salt reactor

1. Introduction

The evaluation of the local cheap natural rocks in the production of road pavement asphalt mixture elements will make a great contribution to the country's economy. The design of aggregate plants, crusher, and sieve selection are important in road construction [1–3] to provide ideal gradation. Formations such as truss, tuff, volcanic slag, and pumice in the regional quarries allow the production of road pavement as aggregate fillers [4, 5]. The pore structure and low mechanical strength cause a negative effect on road pavement quality and avoid the aggregate stone production and capacity [6–9]. For this reason, the texture and microstructure were critical for the quality of mixed asphalt gravel pavement at mechanical resistivity [10–12].

The crushed stone (asphalt gravel, concrete, road material, etc.) used in the construction sector is generally extracted with hammer and hammer-type crushers.

Although this type of crusher produces a higher proportion of dust, it provides advantageous angular grain formation in terms of grain shape [13–15]. However, the crushed road pavement aggregate by hammer caused to excessive cracks during impact breakage. The production of high porous, crack texture was the most suitable for mixing bitumen to easily stick, and that type of aggregate gravel and bitumen mixture should be used in road asphalt. Road pavement aggregate should not be over-broken [16–18].

In Şırnak Province, within the scope of Cizre (45 km away from Şırnak), municipalities produced a total of 1,100,000 tons of aggregate annually by ready-mixed concrete plants and approximately 1,300,000 tons of aggregate ready-mixed concrete [19–21]. The hardness and strength of the aggregate are effective on ready-mixed concrete strength. Therefore, the breaking of the rock by blasting, hammer breaker breakage, sieving, and equipment affects the quality of concrete and road pavement aggregates [18].

Limestone-type aggregates are dolomitic or calcareous rocks and marl, truss, pumice, and basalt-type aggregates are generally alkaline and sometimes contain alkali sulfate and phosphate salts, so they can be partially dissolved by aqueous acidic waste or acidic solutions [22, 23]. The natural pore textures of the medium hard rock such as granular slag, limestone, tuff, marly shale, and asphaltite slime could be mechanically or chemically developed. Macro and micropores of aggregate materials are dependent on the degree of chemical dissolution by atmospherically conditions thaw freeze and rains waters such as acidic mineral waste water contact as cover asphalt mixing use. The creep action of the chemical solution affects the pore structure and mechanical resistivity on road pavement cover [24].

Howard and Datta have determined that chemical grinding has many advantages [25, 26] and that chemical interaction is consistent with grinding of cement raw materials, and bond grindability (G) values are improved by 18% with 0.1 M sulfuric acid and waste acidic coal mine water [23, 24].

Chemical and mineralogical properties of light aggregate rocks should be examined for partial chemical dissolution and pore formation. In this study, chemical acidic interaction and mineralogical parameters of gravel aggregate limestone, marl, marly limestone, basalt, truss, tuff, and marly shale were determined and examined to determine microwave processing of a mixture of asphalt and Şırnak asphaltite slime melting and mixing at certain gradation develop with in compression ability, shear and tension strength and high creep resistivity at appropriate chemical interaction and aggregate type [27].

Limestone, porous limestone, marl, and shale-type light aggregates of Şırnak Province contain heterogeneous structures due to their strength, hardness and porosity properties. Five samples of local road pavement aggregate were investigated in this study. The samples selected in this study were subjected to two stage crushing and sieving process. As a result of these hammer crushing systems called secondary and tertiary materials, the final stage of the crushed stone production has been completed by classifying the material in accordance with the standard of ASTM C330 [28] in the dimensions below 40 mm by passing through the final sieves [29].

Due to the low porosity of marl and marly limestone sample, it is not possible to use it as road pavement aggregate, but it has been examined by evaluating the porosity by chemical solution.

Approximately 90–95% by weight of the ready-mixed concrete mixture and 80–85% by volume constitute road pavement aggregate. Asphalt mixture aggregate type, porosity, grain shape structure, gradation, such as features was investigated on aggregate regarding standard tests affect the use [29, 30]. Boiler base slag in Şırnak Province, which provides the production of semi-light aggregate by chemical interaction, produced road pavement concrete by adding 50% weight ratio (approximately 75–80% volume) The properties required for aggregates to be used in light concrete production are mainly low density and high porosity.

2. Waste rocks and road pavement rocks

In this study, 20 × 5 × 5 cm cubic marl and marl limestone samples were subjected to point-load strength and uniaxial compressive strength tests under laboratory conditions. Macrostructure, microstructure, and mineralogical investigations were conducted [28].

A modified impact shatter test was applied to determine the impact strength due to the porous structure of the road pavement aggregate.

In accordance with ASTM C330 [28], the coarser fraction over No4 sieve and dust fraction of No200 sieves were eliminated from representative aggregate samples, separated, and sieved according to $\sqrt[4]{2}$ and $\sqrt[4]{2}$ series. Samples were prepared using maximum packing density [29].

The aggregate sites that can be used in road construction in Şırnak Province are mainly found in various regions and coal quarries such as: altered limestone (Şırnak Center), marly limestone (Şırnak Center), marl (Şırnak Center), Cizre white porous limestone (Şırnak Cizre), porous limestone, (Cizre stream) volcanic cinder, Midyat limestone, Şırnak coal mine waste marly shale.

The aggregates produced in crushed quarries differ according to each other. These differences are particularly related to the geological structure and the method of operation chosen accordingly. The limestone and marl of Şırnak Province can be used as concrete aggregates due to high mechanical strength. However, the strong parts of this region consist of heterogeneous brownish-yellow limestone formations, showing a variable compression strength of 40 MPa to 70 MPa as altered rock units. This region is preferred for road pavement asphalt mixture aggregates due to the low cost of handling of aggregate from producing quarries.

Today, various industrial wastes are used as concrete pozzolan materials [31–34]. Pozzolans are defined as filler materials that do not have binding properties but which have binding properties when finely ground and react chemically with calcium hydroxide at normal temperature and in humid environments [31–33]. Pozzolans are used as mineral additives in concrete. Mineral additives are used to improve the durability of concrete by increasing the strength [34]. It is known that the addition of a pozzolanic material into concrete or cement provides many benefits, such as reducing the heat of hydration, providing high target strength and low permeability, and controlling the alkali silica reaction and the sulfate effect [35]. The most common pozzolanic materials in the world are silica fume, fly ash, and furnace slag [36]. It is estimated that there are 600 million tons of fly ash in the world today, but only 10% of it is evaluated in concrete and as filler in road pavement covering technology [37]. Fly ash as filler has a wide range of uses in concrete because of lowering the cost of concrete, saving energy, and reducing environmental problems [35]: it is used instead of decreasing the cement in certain proportions, instead of decreasing the fine aggregate in certain proportions, or instead of using both fine aggregate and cement in certain proportions [33–35]. The effects of fly ash on the mechanical properties of pavement were studied extensively. In this study, the effects of Silopi fly ash as a filler additive on hot and hardening cover in road pavement and on the properties of thaw and freeze strength were mentioned, but in this study, the effect of the amount of ash replaced to fine filler amount, change on compression ability, workability, and shear strength of mixture was determined.

Components	Seyitömer fly ash	Şırnak fly ash	ASTM C616 F	ASTM C616 C
SiO_2	54.11	40.71		
Al_2O_3	20.58	11.53		
Fe_2O_3	9.33	5.62		
S + A + F	84.52	88.6	≥70	≥50
CaO	4.72	19.56		
MgO	4.33	2.41	≤5	≤5
SO_4	0.72	1.02	≤5	≤5
K_2O	2.10	2.44		
Na_2O	0.67	0.55		
Ignition loss	3.22	4.74	≤6 ≤ 10(TSE639)	≤6 ≤ 10(TSE639)
Cl	0.006	0.0		
Free CaO	1.72	3.13		
React. SiO_2	39.8	34.12		
React. CaO	2.16	12.72		

Table 1.
Seyitömer and Sirnak Silopi fly ash chemical compositions and aggregate compliance with standards values, EN 450 TS EN 197-1 TS 639 ASTM C616 [37].

The ratio of clay and fly ash greatly affected the water content in the filler material significantly. As the fillers evaluated according to the percentage of clay and fly ash in the samples showed cohesive or low cohesive property. The specific unit weights are given in **Table 1**. Briquetting tests showed higher compressibility for the Şırnak Silopi fly ash, so it gave higher advantageous effect on compaction of road pavement [27].

3. Materials and methods

Mardin Limak composite cement type, CEM IV 42.5 N type cement, was used as a binder. The fly ash used in the experiments was obtained from the Şırnak Silopi Thermal Power Plant, and the chemical composition of the fly ash is given in **Table 2**.

In this study, it is aimed to produce asphalt aggregate mixture briquette blocks at size 10 × 10 × 10 cm^3. In order to produce a high class road cover, bitumen or asphalt/aggregate (B/A) ratio was investigated for high shear strength, creep, and thaw resistivity; the selected aggregates and binder fillers such as carbon or coal slime were added in certain amounts by considering to receive the high compressive strength from asphalt briquettes compressed under 3 tons/cm^2 load. As a result of preliminary experiments, the B/A ratio was decided to be 6–10% and 0.60% polymer and 5–10% Şırnak asphaltite slime or carbon fine. In ASTM standards [28], the amount of mixing bitumen and aggregate rates were taken; however, at the amount of coal and porous aggregate content, amounts of mixture components used in the road pavement cover used for this experimentation are given in **Table 3**.

After the concrete mixture calculations, concrete castings were made to cube molds having dimensions of 10 × 10 × 10 cm for each concrete type. In each series, four batches of concrete were produced, replacing 5, 10, 20, and 30% of the fly ash

Component,%	Şırnak limestone	Sırnak marly limestone	Şırnak marl	Şırnak porous limestone	Şırnak shale
SiO_2	3.53	9.42	24.14	2.12	48.53
Al_2O_3	2.23	6.53	12.61	1.71	24.61
Fe_2O_3	0.59	4.48	7.34	0.58	7.59
CaO	49.48	39.23	29.18	45.22	9.48
MgO	2.20	2.28	4.68	7.41	3.28
K_2O	0.41	0.53	3.32	0.40	2.51
Na_2O	0.35	0.24	1.11	0.21	0.35
Ignition loss	46.19	26.11	21.43	48.04	3.09
SO_3	0.32	0.21	0.20	0.02	0.32
Density	2.71	2.68	2.55	2.33	2.42

Table 2.
The chemical analysis of aggregate mixture fillers of Şırnak Province limestone, marl, and shale.

Component,%	Şırnak boiler bottom ash-slag	Volcanic slag	Tatvan pumice	Şırnak fly ash
SiO_2	43.48	50.50	60.13	41.48
Al_2O_3	16.10	14.61	17.22	18.10
Fe_2O_3	10.52	24.30	4.59	4.52
CaO	8.48	2.30	2.48	18.48
MgO	3.80	1.28	2.17	4.20
K_2O	2.51	2.51	3.51	2.71
Na_2O	1.35	1.35	4.35	1.95
Ignition loss	10.9	0.21	4.12	1.9
SO_3	0.32	0.12	0.52	0.22

Table 3.
Chemical composition and binder fillers for bitumen aggregate present in the Şırnak regional waste aggregate samples.

cement. To obtain reference values, a series of normal concrete block is produced without fly ash. The concrete workability of the produced concrete series was determined by shear box test and shatter experiments. For each prepared mix, the compressibility of the asphalt aggregate briquettes compressed with 3 tons/cm^2 was managed under loading 50, 100, and 500 MPa which was calculated by deformation level percentage. The shear resistivity was tested using the method of penetration resistance. Steel cone is indented over the briquette and penetration diameter calculated as shear resistivity under 50 MPa. In addition, the shear box test used for briquettes and shatter tests was carried out for shatter strength of the asphalt aggregate briquettes obtained from the prepared mixtures of Şırnak asphaltite slime, fly ash, and polymer.

There are a number of experimental methods that proposed so far that bitumen aggregate can be homogenized by heating over 100°C to set the aggregate and asphalt mixing like dough. There are different test methods to determine the quality of asphalt and mixed polymer-based components on measuring the shear of the bitumen aggregate mixtures, compressibility by the vibration, heating, hardness change at the cooled temperature following the solidification, change in compression volume, and resistance of the briquette against penetration [29–47].

3.1 Chemical analysis and bright section analysis

Mineralogical compositions of the samples were determined by means of standard chemical Ca, Mg, and silica analyses. The samples were first brought to dimensions between 40 and 10 mm in jaw crusher and were homogenized by milling to 0.1 mm. Powder samples are thawed and burned with HF in a platinum crucible for silica content. Chemical composition of the rocks provided in the vicinity of Şırnak Province in the experiments is given in **Table 1**. The amount of silica in the marly and marly limestone was reduced.

Prior to the preparation of the bright section, a yellow liquid epoxy resin was impregnated with the samples in a medium vacuum. This resin penetrates into the pores and makes the pores appear easier under the microscope. The porous texture is seen in **Figure 1** for different types of aggregates.

3.2 Effect of chemical interaction on aggregate

Chemical resistivity of grains was determined by contact of aggregates with chemically acidic waters and acidic mineral waters at a certain period of time. At the end of contact with the solutions, the aggregates dissolved at a certain level regarding their carbonates and salt content and became porous by solution. This makes it possible for the original chemical structure of the original aggregate type

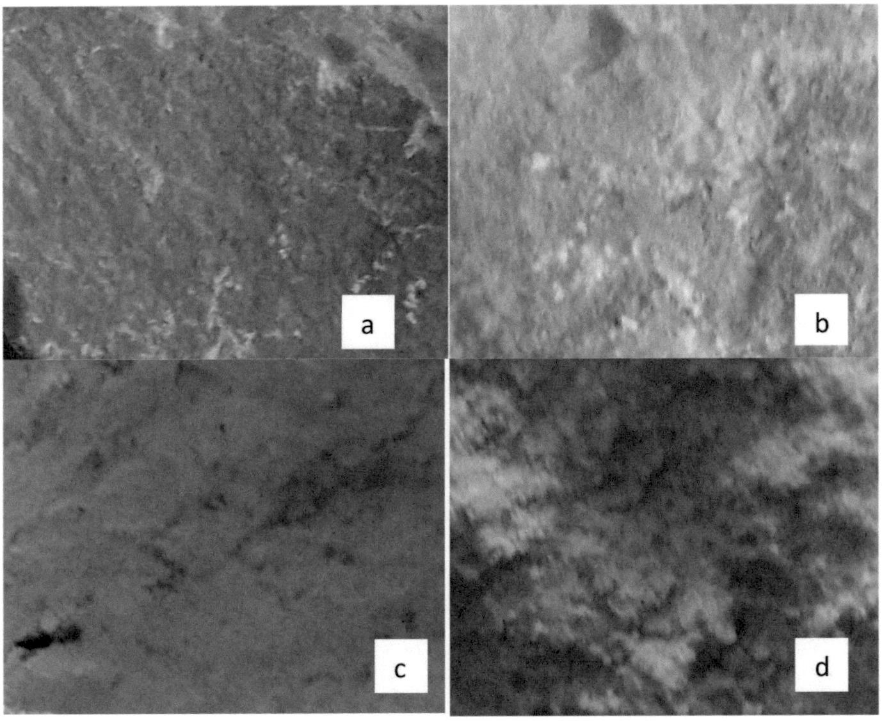

Figure 1.
(a). Marly shale, (b) Şırnak limestone, (c) Şırnak porous limestone, and (d) the chemically degraded porous limestone.

to be limestone. For this purpose, different types of aggregates were stored as 1–2 kg aggregate samples in 20 lt acid containers for 1–3 days to evaluate their effect on various aggregates in the treatment with 50–70°C 1 M HCl and 1 M H_2SO_4 acid. The products were washed to determine the weight change and apparent specificity and pore exchange with the calculation. The apparent specific gravity variation of the aggregate bulk in the visible container according to the chemical interaction is shown in **Figure 2**.

The interaction effect is shown in **Figure 2** for limestone and marl. After the chemical interaction, the limestone aggregates and porous limestone aggregates for the used asphalt mixture were tested for the modified impact loading and shatter test. Strength of asphalt aggregate mixtures was measuring the shear resistivity by cycling loads as for light concrete blocks were produced at lower compression strength and bending strengths [7–9]. Bulk unit weight, bulk void content, void percentage, and noncompact briquette strength relationships were excluded. Optimum strength for asphalt aggregate mixture and fly ash aggregate mixtures and change of amount of aggregate content were determined.

3.3 Grain size analysis

Representative samples of approximately 1–0.5 kg were screened with standard sieve series for 20 min (ASTM C136) [30]. No4 and No200 sieves in accordance with ASTM C330 [28] standard sieves were sieved and the samples were separated according to $\sqrt{2}$ and $4\sqrt{2}$ series, and size distributions were determined according to Gaudin-Schumann and RRS theory. The results of sieve analysis of rocks are given in **Table 4**. **Figure 3** shows fine filler material size distribution and **Figures 4** and **5** illustrate aggregate distribution in Gaudin-Schumann and RRS theory, respectively.

During these tests, Gaudin-Schumann distribution index was examined for road pavement concrete production. The distribution was determined as % s/mm. According to these tests, it was observed that the Gaudin-Schuman grain size approach provides a more suitable apparent density than RRS.

The void ratio of 40 mm maximum grain size was determined as 1/1 + v, pore%: 1 – (1/1 + v)] by 1% roughness measurement, and (1/1 + v)% was solid volume occupancy rate with pore% calculation.

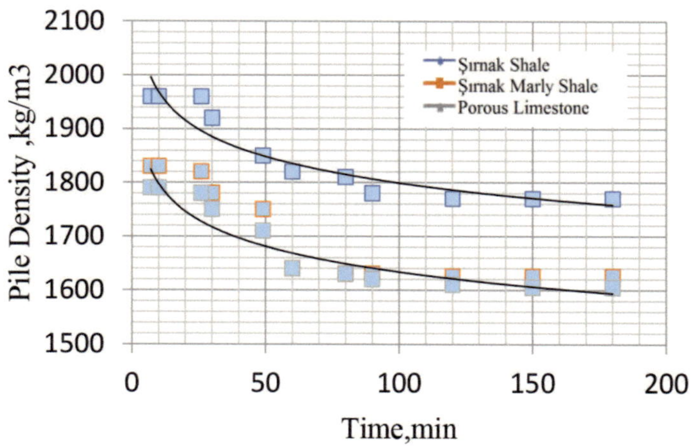

Figure 2.
Alteration resistivity by density change measurements of aggregate specimens hold in acidic solutions Şırnak marl (blue), Şırnak limestone (yellow), and Şırnak marly shale.

Aggregate type	Sieve aperture, mm	25	16	10	5	2.8	1.2	0.4	0.2
Şırnak shale	0–5				99	95	76	27	12
	5–15		100	48	0.2				
	15–40	98	3						
Sırnak marl	0–5				99	87	66	21	9
	5–15		100	38	0.2				
	15–40	98	2						
Şırnak porous limestone	0–5				99	79	60	19	7
	5–15		100	28	0.2				
	15–40	96	1						
Midyat limestone	0–5				99	77	54	17	6
	5–15		100	24	0.0				
	15–40	94	0.2						
Şırnak porous limestone	0–5				99	76	52	16	5
	5–15		100	23	0.0				
	15–40	93	0.2						
	5–15		100	21	0.0				
	15–40	88	0.1						

Table 4.
The sieve analysis of the aggregates used in this study.

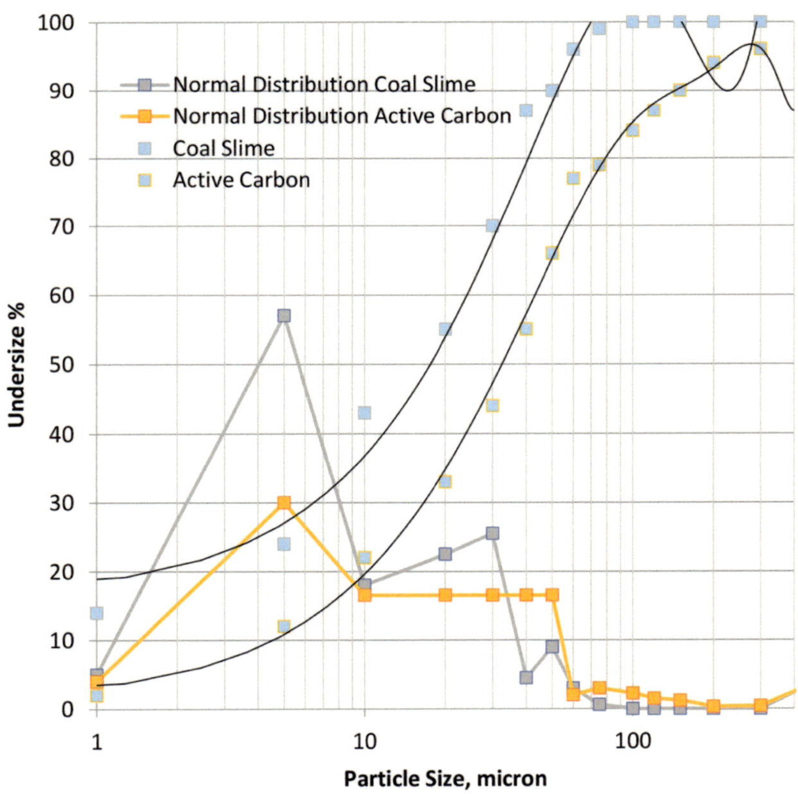

Figure 3.
Aggregate surface tension grading fillers for polymer and asphalt mixtures in Şırnak.

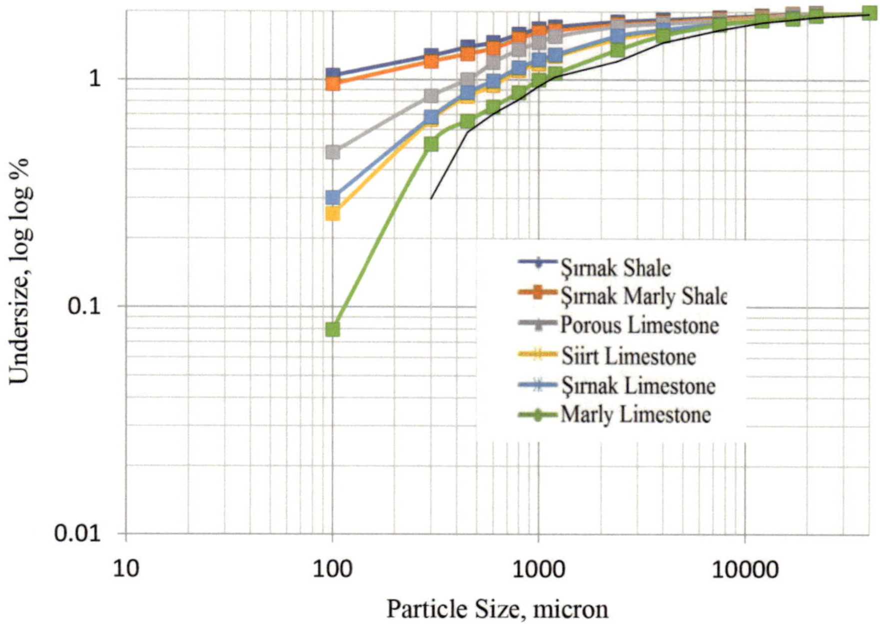

Figure 4.
Variation of Gaudin-Schumann particle size distribution and normal size distribution.

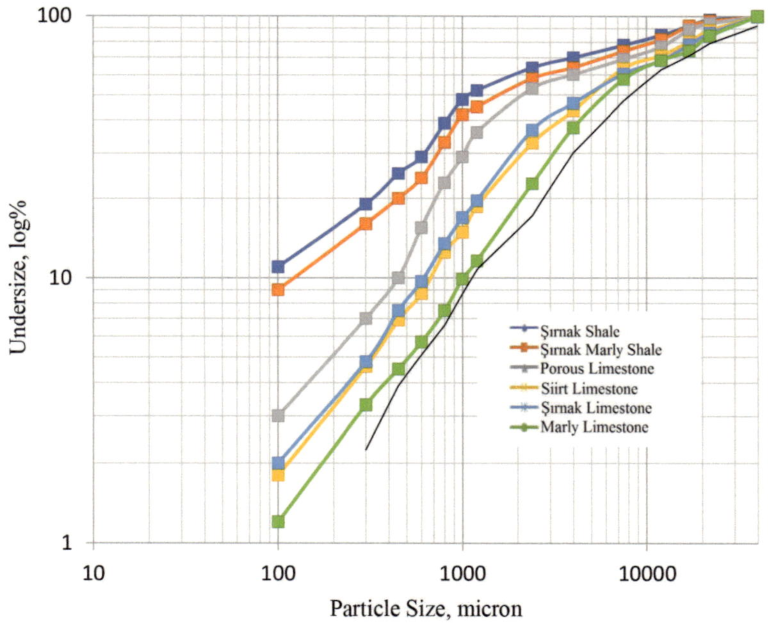

Figure 5.
Variation of RRS particle size distribution and normal size distribution of aggregates.

4. Aggregate asphalt mixture briquette stability

Regarding compression, the natural stone ground in gravel size fractions of 25 10 and over 25–36 mm is compressed hardly, according to the compression parameters in briquetting tests, hammer breaker crushed aggregate stones sieved for fine dusty particles and slightly compacted in a litter box packed again. In this case, the stability of compacted aggregate parameters such as dry density and porosity is used in the analysis. Four separate compacted block samples were taken from different briquettes of the aggregate mixture, and cutting box test was carried out to determine the shear strength parameters prior to asphalt mixing and compaction. After the experiments were made, the Mohr Coulomb criteria for compacted aggregates gave the c′ and ɸ′ values, and compression strength of briquettes σ was determined. **Figure 6** shows the deformation tension stress chart.

The elder limestone in the province of Şırnak showed a higher uniaxial compression strength of 93 MPa with high silica content, and 24–56 micron sized grains provided suitable conformity with the solid concrete specifications and even road pavement. The Midyat limestone located in Mardin Province, Urfa, Batman, Siirt region, has large potential reserves, covers a large area, and has a very porous texture with 12–14% pore content ranging in macro size 500–3 mm and therefore was very suitable for road pavement aggregate. However, the creek edges in Şırnak Province may contain Mesozoic aged sandstone-clayish schist called Siirt formation. It shows a more homogeneous structure. Especially, the mixing of clayey schists into the aggregate produced can cause problems in terms of plasticity and pore of light aggregates [4–6].

The calcareous cavities and marly formations in the late Mesozoic limestone in the Kasrik region of Şırnak are a great source for filling aggregate for road pavement on the contrary that the aggregate producers should focus on river pebble excavation from river bed due to the lower operation cost. Moreover, the water absorption rates of the aggregates in this region between 2.7 and 3.1% are advantageous in the production of road pavement concrete.

In the area where dolomitic limestones are generally exposed, the limestones are exposed to the fractures formed as a result of faulting. In the areas of aggregate production, marly shale and disseminated brown clayish muddy loose formations can be observed. There is a significant gradation problem in light aggregate production.

The mechanical tests were made related to ASTM rock standards and given in **Table 5**.

Şırnak limestone, Midyat limestone, Şırnak marl, and marly shale were sieved and −15 mm faction were compacted with fly ash under 3 ton compression The blocks were tested after certain curing time and dried in air temperature.

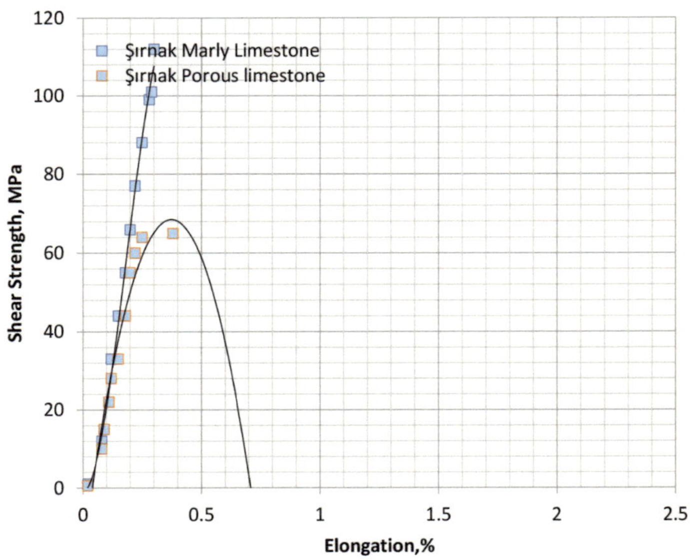

Figure 6.
The Mohr Coulomb criteria for compacted aggregate blocks by Şırnak Silopi fly ash in 14 days.

Aggregate type	Compression strength, MPa	Cohesion, MPa	Internal friction	Shear strength	Shear module
Şırnak shale	11	1	22	22	0.2
	9.8	0.8	33	33	0.3
	11.7	0.7	26	26	0.25
Sırnak marl	11	1.1	22	22	0.2
	9.8	0.8	33	33	0.3
	11.7	0.7	26	26	0.25
Şırnak porous limestone	11	1	22	22	0.2
	9.8	0.8	33	33	0.3
	11.7	0.7	26	26	0.25
Midyat limestone	11	1	22	22	0.2
	9.8	0.8	33	33	0.3
	11.7	0.7	26	26	0.25
Şırnak porous limestone	11	1	22	22	0.2
	9.8	0.8	33	33	0.3
	11.7	0.7	26	26	0.25

Table 5.
Mechanical properties of Şırnak aggregates.

Figure 7.
Uniaxial pressure tester.

Briquettes of aggregates of three different sizes were obtained from limestone and tested in uniaxial compressive strength as seen in **Figure 7**. Aggregate mixing ratios are 2–3 mm 20% natural shale, 30% sieved representative −5 mm crushed stone, 25% sieved fraction of +5 to 16 mm crushed stone, and 25% + 16 to 25 mm, with the largest grain size 32 mm and within the limit values given in TS 706 set as crushed stone. The results of sieve analysis of the aggregates used are given in **Tables 6–8**.

Curing time min	Uniaxial compressive strength, MPa				
Compacted binder	1. Sample	2. Sample	3. Sample	4. Sample	Average sample
0.5% fly ash					
	192.56	192.56	184	180	185.52
1	205.6	205.6	209.60	203.5	204.55
3	284.5	284.5	288.4	296.5	289.80
5% fly ash					182.99
0.5	192.56	183.1	180.06	185.81	185.52
1	205.6	188.81	198.39	196.1	194.43
3	284.5	304.84	319.7	301.75	308.76
10% fly ash					203.61
0.5	192.56	200.99	206.99	202.86	185.52
1	205.6	251.17	253.9	257.56	254.21
3	284.5	360.79	369.2	360.73	363.57
20% fly ash					205.50
0.5	192.56	232.86	170.49	213.15	185.52
1	205.6	300.24	220.92	241.46	254.21
3	284.5	355.13	356.99	359.21	357.11
25% fly ash					281.89
0.5	192.56	290.24	280.24	275.2	185.52
1	205.6	323.66	316.99	313.72	318.12
3	284.5	381.17	377.42	371.8	376.80

Table 6.
Compression strength of aggregate asphalt mixture of aggregate type of Şırnak limestone.

4.1 Creep characterization of asphalt mixture limestone marl and shale: thaw-drop sequential ball impact strength of porous aggregate

The chemical activation of limestone, marly limestone, shale, and marl produced from Şırnak coal mines and quarries was investigated, and the qualities of products of semi-road pavement aggregates providing the specifications required from road pavement concrete production were determined. The waste aggregates were treated with 1 M HCl and 1 M H_2SO_4 acidic solutions in order to produce a porous aggregate. A 3–4 day treatment has become sufficient. A temperature of 50°C increased activation. The compression strengths of road pavement concrete blocks varied between 7.2 and 18.4 MPa. Modified impact shatter resistance was also discussed, hence the strengths.

The semi-graded aggregate samples were subjected to compressive strength and impact shatter strength test and the size distribution and impact mechanical strength of the crushed product were determined (ASTM D6024-07) [29]. The classified aggregate samples were examined as the size of the fixed 5 kg impact sledgehammer weight of 10 × 10 × 10 cm^3 with the effect of 50 cm falling in the hive, and the product was compared to the light concrete samples. With these technological applications, the evaluation of local natural stones as a road pavement building aggregate can be further improved. In the impact shatter strength tests, 5 kg sledge weight was taken from four times 50 cm height to fall on the 40, 25 mm aggregate inside the 8 cm inner diameter steel shank, under 5 mm less and 25 mm above the weight as hard, the strength was higher. The results are shown in **Figure 8**.

Curing time min	Uniaxial compressive strength, MPa				
Compacted binder	**1. Sample**	**2. Sample**	**3. Sample**	**4. Sample**	**Average sample**
0.5% fly ash					
	192.56	192.56	184	180	185.52
1	205.6	205.6	209.60	203.5	204.55
3	284.5	284.5	288.4	296.5	289.80
5% fly ash					182.99
0.5	192.56	183.1	180.06	185.81	185.52
1	205.6	188.81	198.39	196.1	194.43
3	284.5	304.84	319.7	301.75	308.76
10% fly ash					203.61
0.5	192.56	200.99	206.99	202.86	185.52
1	205.6	251.17	253.9	257.56	254.21
3	284.5	360.79	369.2	360.73	363.57
20% fly ash					205.50
0.5	192.56	232.86	170.49	213.15	185.52
1	205.6	300.24	220.92	241.46	254.21
3	284.5	355.13	356.99	359.21	357.11
25% fly ash					281.89
0.5	192.56	290.24	280.24	275.2	185.52
1	205.6	323.66	316.99	313.72	318.12
3	284.5	381.17	377.42	371.8	376.80

Table 7.
Compression strength of aggregate asphalt mixture of aggregate of Şırnak marl.

4.2 Thaw-freeze drying creep impact resistance tests

Concrete aggregate samples were subjected to a cyclic modified creep test and the product has a concrete aggregate strength of 5 kg of steel ball impact mechanical strength (ASTM D6024) [29]. The classified aggregate samples were examined as the dusting of the size of the fixed 5 kg impact sledge hammer with the effect of 8 × 50 × 50 cm falling in the hive, and the product was compared with the concrete compressive strength. With these technological applications, the evaluation of local natural stones as a road pavement building aggregate can be further improved. In impact resistance tests, 5 kg sledge weight was taken from four times 50 cm height to fall on 50 × 50 × 25 mm like block aggregate inside the inner diameter of 8 cm inner diameter steel sleeve. The results are shown in **Figure 8**. Grain shape replacements with photographic technique assumed that there is deformation, calculated using close volume changes.

4.3 Compaction of bitumen composite and cracking tests

Compaction was carried out over 30 ton load over 5 cm thick samples, and microscopy evaluation of local aggregate stones as a road bitumen is shown in **Figures 9–12** the crack forms on additive hardener. Further polymer addition at 5% rate improved the microstructure to micronized grains. Impact shatter test resistance tests provided similar results as given in the next discussion.

Curing time min	Uniaxial compressive strength, MPa				
Compacted binder	**1. Sample**	**2. Sample**	**3. Sample**	**4. Sample**	**Average sample**
0.5% fly ash					
	192.56	192.56	184	180	185.52
1	205.6	205.6	209.60	203.5	204.55
3	284.5	284.5	288.4	296.5	289.80
5% fly ash					182.99
0.5	192.56	183.1	180.06	185.81	185.52
1	205.6	188.81	198.39	196.1	194.43
3	284.5	304.84	319.7	301.75	308.76
10% fly ash					203.61
0.5	192.56	200.99	206.99	202.86	185.52
1	205.6	251.17	253.9	257.56	254.21
3	284.5	360.79	369.2	360.73	363.57
20% fly ash					205.50
0.5	192.56	232.86	170.49	213.15	185.52
1	205.6	300.24	220.92	241.46	254.21
3	284.5	355.13	356.99	359.21	357.11
25% fly ash					281.89
0.5	192.56	290.24	280.24	275.2	185.52
1	205.6	323.66	316.99	313.72	318.12
3	284.5	381.17	377.42	371.8	376.80

Table 8.
Compression strength of aggregate asphalt mixture of aggregate of Şırnak porous limestone.

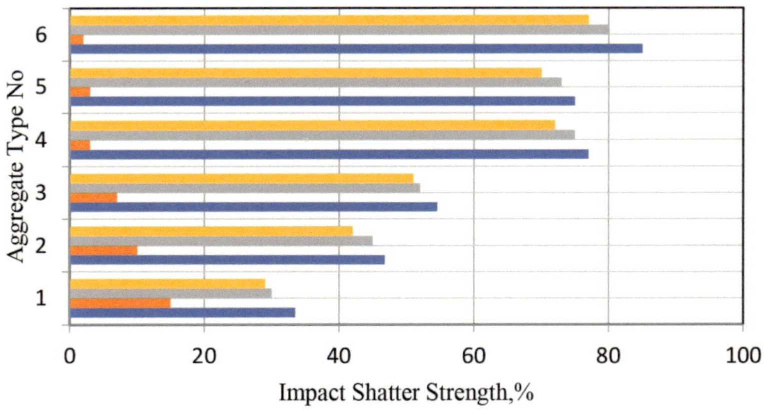

Figure 8.
Histogram view of impact, abrasion resistance, 1 M H_2SO_4 and 1 M HCl at 3 days of the 7% bitumen aggregate briquettes; (1). Sırnak porous limestone, (2) Sırnak marl, (3) Şırnak marly shale, (4) Midyat limestone, (5) Şırnak porous limestone, and (6) Şırnak marl.

The asphaltite bitumen composite cracking results are shown in **Figure 10**. Grain shape and volume replacements with photographic technique were tested assuming that there is deformation, calculated using density changes. The density change by addition and distribution of homogenized bitumen in thin sections is shown in **Figure 11**.

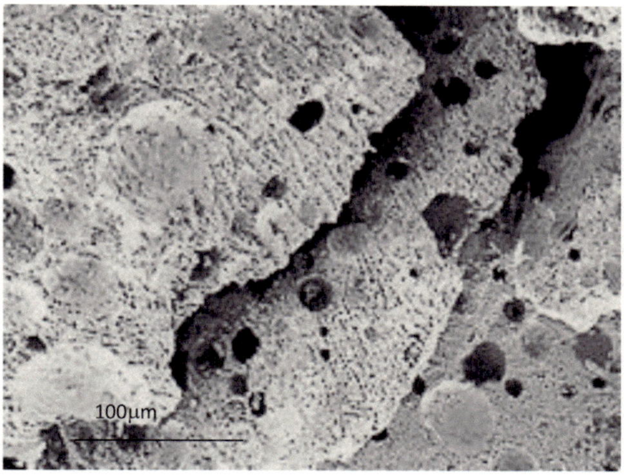

Figure 9.
Asphaltite slime and shale clay aggregate compressed briquette.

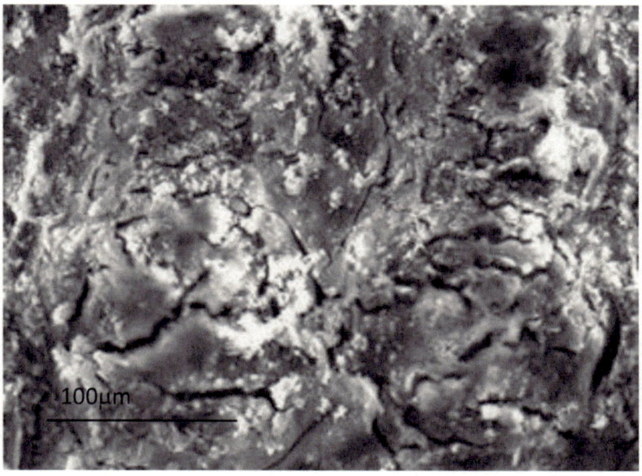

Figure 10.
Asphalt—Şırnak asphaltite and Şırnak marly shale aggregate compressed briquette.

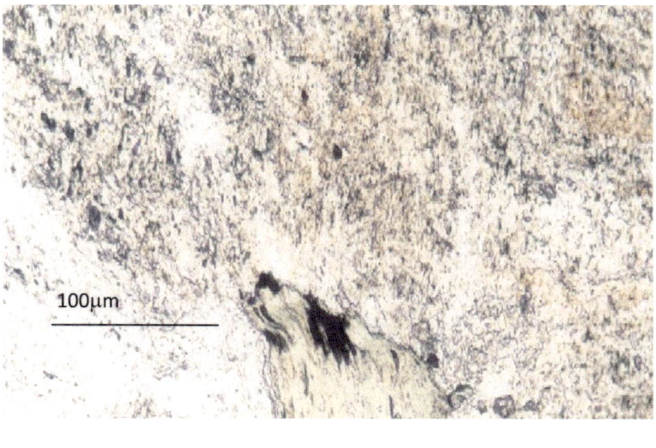

Figure 11.
Asphalt mixed Şırnak asphaltite shale thin section.

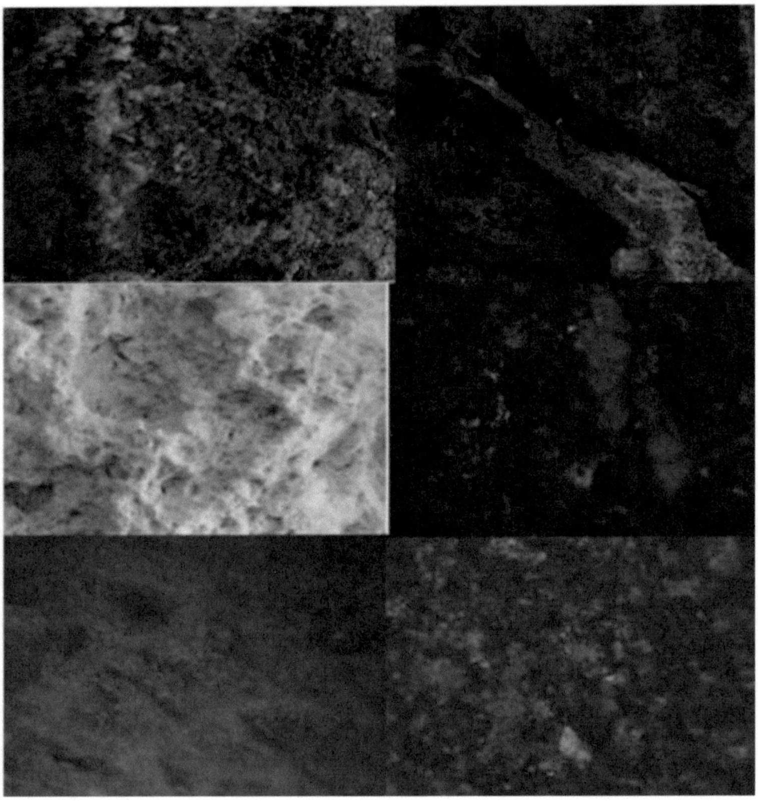

Figure 12.
Microchange of bitumen compression on compression test.

The cracking on compaction overload of 30 ton is shown in **Figure 12**, and crack occurrence was odd with higher breakage and no bitumen contact in cracks.

5. Results and discussions

5.1 Tension loading and compressibility shear tests of bitumen mixtures

The high-graded bitumen samples were subjected to tension, shear, and compressive strength and tension meter deformation test, and the size distortion was determined (ASTM D6024-07). The classified asphaltite slime and bitumen polymer acrylate samples were homogenized at 110°C in a microwave oven for 10 min and the size of the fixed thin plate of 5 × 20 × 100 mm^3 was examined with the effect of 1 ton loading cm in the tension meter ring, and the final elongation and broken state was compared to the start-up state. With these standard applications, the evaluation of local composite bitumen as a road pavement aggregate can be further improved. Test results are shown in **Figure 13** for the tension meter ring and tension stress through deformation. The tests of shear resistance were made in a shear box of 50 mm length ring. The results are shown in **Figure 14**. Shear distance changed over the 250 mm the compressibility with the 3 cm inner diameter steel barrel, under 25 mm swelling compression stress as hard, the strength was higher. The results are shown in **Figure 15**.

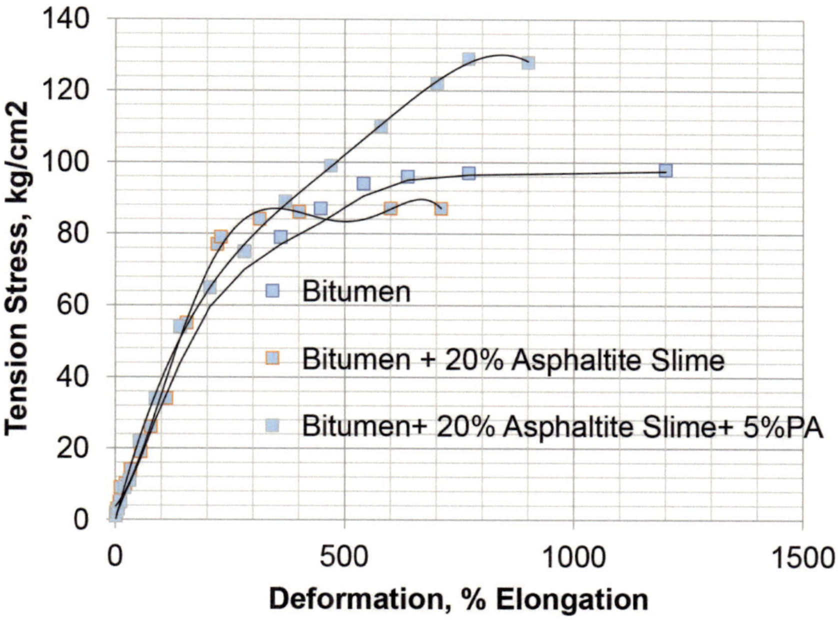

Figure 13.
Deformation change with tension stress test results of wire deformation test rig.

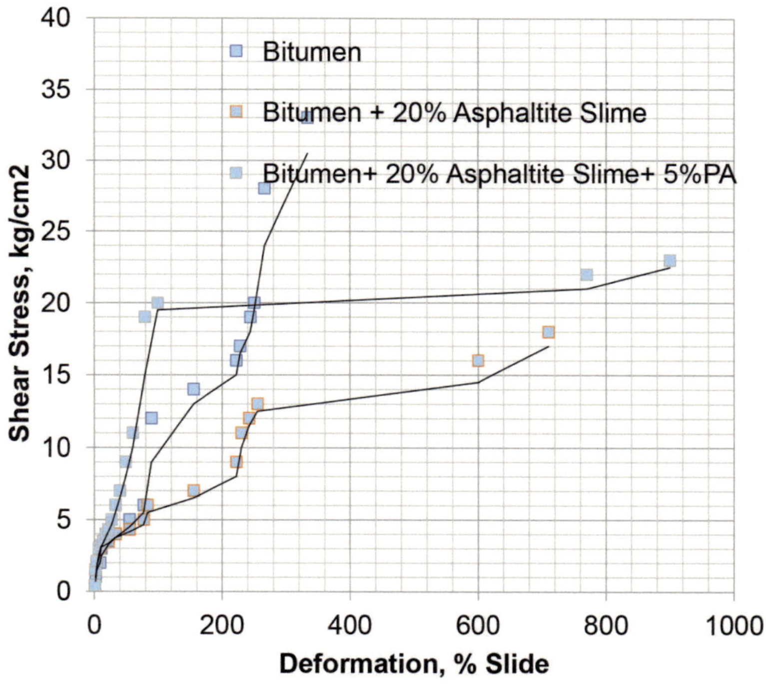

Figure 14.
Shear deformation change with shear box test results of wire deformation test rig.

5.2 Point loading and compressive strength tests of aggregates

The test samples were produced as 5 × 5 × 5 cm blocks and 10 samples were determined to show 95% accuracy rate by prestigious ELE brand. The results are shown in **Figures 16** and **17**.

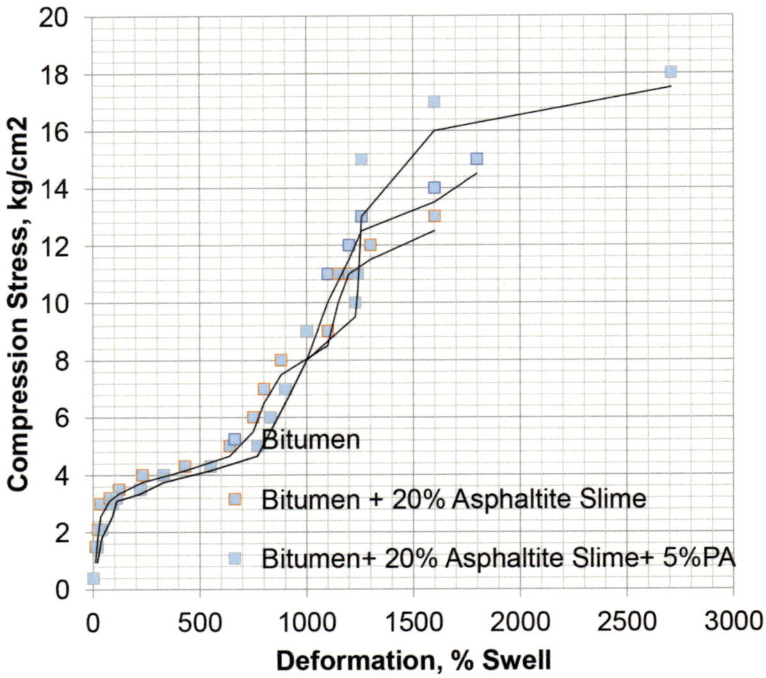

Figure 15.
Compressibility deformation change with compression stress test results of wire deformation test rig.

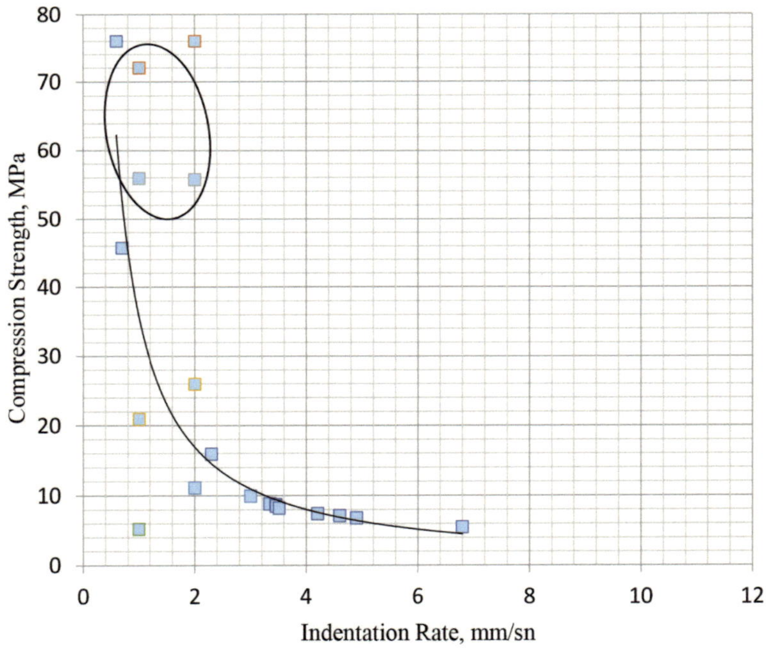

Figure 16.
Compression strength change of the Şırnak aggregates.

Chemically treated samples have also increased the chemical impact of small grain sizes. Approximately, the aggregate in 5 mm contained 13.6% of the pore in bulk pile as 25 mm fraction showed a pore rate as lower at 9.2%. The results are shown in **Figure 2**. **Figure 2** also proved that the chemical interaction of 2 days is sufficient. Hence, the dissolution process reached the alkaline lime salt saturation of the solution. The pores in the limestone sample reached 13.6%.

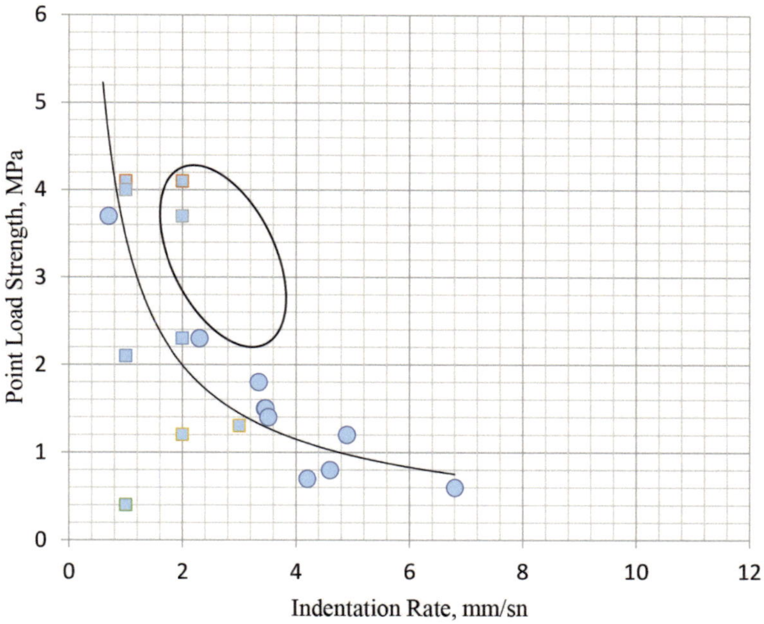

Figure 17.
Point load strength change of the Şırnak aggregates.

Porous limestone texture, chemical interaction result, and petrographic changes are seen in **Figure 2**. It was determined that the amount of silicate of 2, 3, 3, and 4% of Şırnak altered limestone was changed and the pore size was at a macro size of 1–3 mm and partly micro sized of 50–300 micron. The pore rate was in microcrystalline microporous structure of 13.4–14.8% silica in Şırnak marly limestone and as 5–30 micron micro porous. The degree of chemical interaction in marinated limestone and marls was not sufficient due to the silica content and the microcrystalline pore structure.

The ideal gradation according to **Figures 4** and **5** can be determined by the Gaudin-Schumann approach. The distribution factor in the marly limestone was increased by 81%, and in Shale, it decreased to 45%. The effect of chemical interaction time was tested. The treatments of 3 days have increased. As shown in **Figure 2**, impact shatter strength values were decreased by about 24% over a two-day period. The pore ratio was also similar in the limestone and shale samples. It was increased by 17%. It was caused by the similarity of the original pore sizes and grain structures of the structural structures. In addition, the calcite and dolomite component within the shale stone provided this. The results did not change in 2 and 3 days. In these evaluations, the values obtained in the impact resistance test are shown in **Figure 8**.

The homogeneous fine-grained structure has increased the strength even more. The heterogeneous silicate component, especially in Şırnak, has a high impact resistance. Since the silica content of Şırnak and Siirt limestone is low, they are soft and have less impact and abrasion resistance.

Figure 8 shows the resistance values. Limestone samples vary depending on the amount of silicate they contain, and also have changed its microstructural properties and related to the pore. Wear conditions of the pore density in the development of abrasion and consequently the formation of brittle surfaces and lower wear resistance values were observed.

Although the Şırnak marl limestone and marllyleyl rocks are of the same type, they have different wear and strength properties. Regarding impact shatter tests, the impact strength of the limestones of Şırnak is higher than that of the other samples. The reason for this is their porosity and clay texture.

5.3 Bituminous aggregate mixture block production

In this study, 50% weight of Şırnak Province with 2% chemically interacted limestone aggregate and 50% weight ratio of Şırnak Kazan slags were evaluated as light aggregate and road pavement concrete production. Slag samples were crushed to size 40 mm below the aggregate [6, 7, 13]. Aggregate mixing ratios were below 28 mm. The bituminous mixture aggregate (% 50 coarse aggregate), fine aggregate (% 50 fine sand) and% 25 is set to bituminous mixture. In this study, it is aimed to produce C 20 class aggregate bituminous polymer asphaltite slime homogenized at 90°C. Coal tar as bitumen was used as a binder. Polymer/bitumen (p/B) ratio was critical and pre-determined to be 0.30 as a result of preliminary tests. The compressive strength of bıtumen asphaltite and polymer rods extruded at 3 ton with different aggregate types and cooled to ambient temperature was determined as shown in **Figure 18**.

In the uniaxial compressive strength tests, the highest strength values were determined to vary between 76 and 35 MPa depending on the type of limestone and the silicate compositions. The plume ratio with low uniaxial compressive strength exceeds 5%. Midyat limestone with more than 4% silicate phases was observed. Şırnak porous limestone was found to be 39.3 MPa with a low density of 3.6% in porous matrix and 34.6 MPa with 7.4% porous matrix.

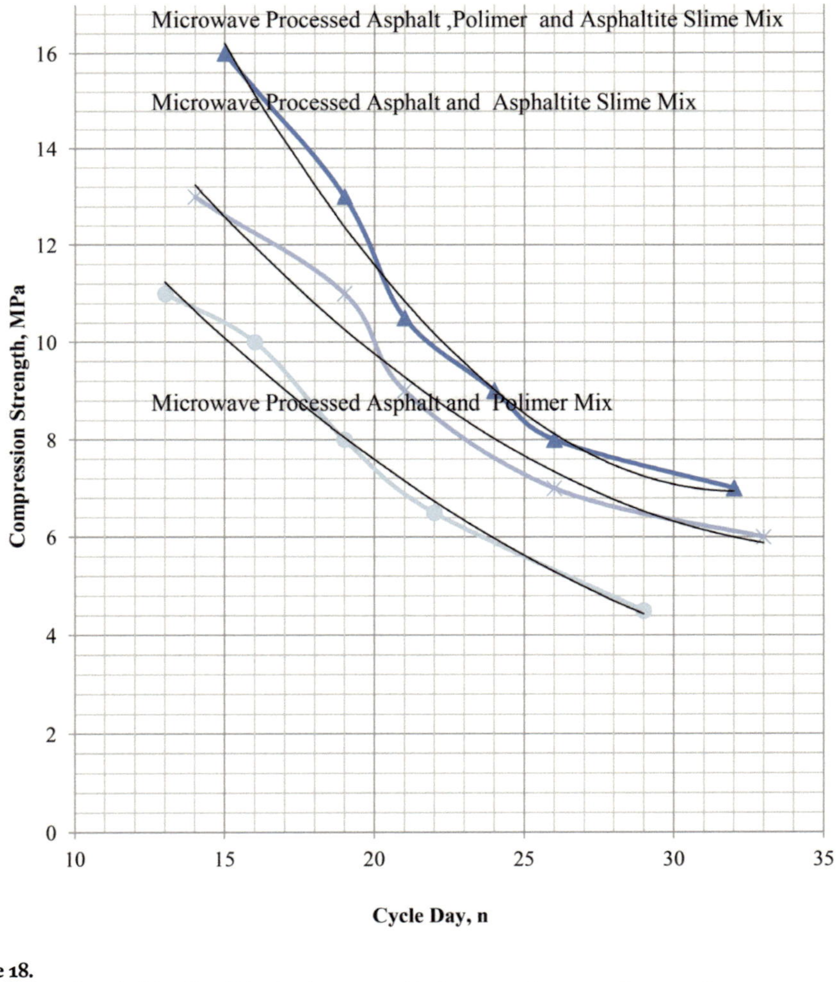

Figure 18.
Bitumen-Şırnak asphaltite slime PA mixed aggregate block compression strength test results.

In **Figure 16**, a similar change was observed when the compressive strengths of different pore and coarse density limestones were examined. Clearly, when the stress-strain diagram is examined in uniaxial compressive strength tests, it is seen that marls and limestone exhibit high strength and this is caused by 0.5–1.8% porosity.

As seen from the point load strength values of **Figure 17**, the coarse grained structure was weaker and more brittle than the marl matrix, and heterogeneous hard marls showed higher compressive strength.

As shown in **Figure 18**, point creep load strength values, the coarse grained structure, marl concrete which is most affected by frost and frost cycle is weaker and more fragile and heterogeneous hard marl has higher compressive strength.

As shown in the strength values of the aggregate concrete blocks produced in **Figure 9**, the structure, chemical interaction, and aggregate concrete produced made the concrete more durable. Heterogenous texture of hard marl reached higher pore. This study showed that the chemical interaction in the altered limestone stopped with saturation and was insufficient in the production of solid blocks.

6. Conclusions

At least the gradation of quarries and the processing of limestone will be of great benefit in the production of porous aggregated bulk.

The most important factor in the formation of a gapless structure in concrete is the regular surface contact and aggregation between the aggregates, that is, in order to fill the gap between coarse aggregates, it is necessary to use technologies that will produce porous light fine aggregates in different ratios (depending on the properties of the fine aggregate). It is very important to use fine aggregates at optimum levels in pavement cover.

In chemical interaction, the amount of pore in the sulfuric acid solution for 3 days increased by 27–37%. It produced about 0.190 kg/lt weight loss, so highly compressible aggregate production.

When the compressive strength results were examined, it was seen that as the pore amount increased, the compressive strength values did not increase the strength of the aggregate.

Light-weight concrete blocks with high-strength slag and road pavement concrete blocks have high strength values, although the impact strength values can be low.

According to these tests, the Gaudin-Schumann particle size approach was found to provide a more favorable approach and tight packing density than the RRS. Certain large and fine-sized aggregate samples are produced with more permeable and porous block asphalt cover.

The highest strength values are controlled by the general texture and micro-structural properties of the aggregate.

It has been observed that shear strength yields successful road pavement cover results in asphalt/asphaltite slime/Şırnak porous limestone aggregate mixing at a weight rate of 8/8/88% briquette compared with truss and volcanic cinder. It was determined that the impact strength values were directly related to the compression resistance of road pavement blocks.

Author details

Yıldırım İsmail Tosun
Faculty of Engineering, Department of Mining Engineering, Şırnak University, Şırnak, Turkey

*Address all correspondence to: yildirimismailtosun@gmail.com

References

[1] Airey GD, Collop AC. Mechanical and structural assessment of laboratory- and field-compacted asphalt mixtures. International Journal of Pavement Engineering. 2014;**17**(1):1-14. DOI: 10.1080/10298436.2014.925551

[2] Tashman L, Wang LB, Thyagarajan S. Microstructure characterization for modeling HMA behavior using imaging technology. Journal of Road Materials and Pavement Design. 2007;**8**(2):207-238

[3] Bieniawski ZT. Mechanism of brittle failure of rock. Part I—Theory of fracture process. International Journal of Rock Mechanics and Mining Science and Geomechanics Abstracts. 1967;**4**(4):395-406

[4] Campione G, La Mendola L. Behavior in compression of lightweight fiber reinforced concrete confined with transverse steel reinforcement. Cement and Concrete Composites. 2004;**26**:645-656

[5] Cavaleri L, Miraglia N, Papia M. Pumice concrete for structural wall panels. Engineering Structures. 2003;**25**:115-125

[6] Chen B, Liu J. Experimental application of mineral admixtures in lightweight concrete with high strength and workability. Construction and Building Materials. 2008;**22**:655-659

[7] Demirboğa R, Orung I, Gül R. Effects of expanded perlite aggregate and mineral admixtures on the compressive strength of low-density concretes. Cement and Concrete Research. 2001; **31**:1627-1632

[8] Demirdag S, Gündüz L. Strength properties of volcanic slag aggregate lightweight concrete for high performance masonry units. Construction and Building Materials. 2008;**22**:135-142

[9] Doel A. Lightweight aggregates for use in concrete. Concrete. 2007;**41**(7):36-37

[10] Emery JJ, Schenk W, Carrick JJ, Davidson JK, MacInnis WK, Kennepohl JA. Stone Mastic Asphalt Trials in Ontario. Washington, DC: Nation Research Council; 1993. pp. 47-53. Transportation Research Record 1427, TRB

[11] Hunter AE, Airey GD, Collop AC. Aggregate orientation and segregation in laboratory compacted asphalt samples. In: Proceedings of the 83rd Annual Meeting. Washington, DC: Transportation Research Board; 2004

[12] Kose S, Guler M, Bahia HU, Massad E. Distribution of strains within asphalt binders in HMA using imaging and finite element techniques. Journal of the Transportation Research Board. 2000;**1728**:21-27

[13] Erdoğan TY. Beton. Ankara: ODTÜ Geliştirme Vakfı Yayıncılık ve İletişim A.Ş.; 2003

[14] Gündüz L. Use of quartet blends containing fly ash, scoria, perlitic pumice and cement to produce cellular hollow lightweight masonry blocks form on-load bearing walls. Construction and Building Materials. 2008;**22**:747-754

[15] Gündüz L. The effects of pumice aggregate/cementations on the low-strength concrete properties. Construction and Building Materials. 2008;**22**(5):721-728

[16] Gündüz L, Bekar M, Şapcı N. Influence of a new type of additive on the performance of polymer-lightweight mortar composites. Cement and Concrete Composites. 2007;**29**:594-602

[17] Gündüz L, Uğur İ. The effects of different fine and coarse pumice aggregate/cementations on the structural concrete properties without using any admixtures. Cement and Concrete Research. 2005;**35**(9):1859-1864

[18] Gündüz L, Sarıısık A, Tozaçan B, Davraz M, Uğur İ, Çankıran O. Pumice Technology. Vol. 1. Isparta: Süleyman Demirel University; 1998. pp. 275-285

[19] Tosun Yİ, Cevizci H, Ceylan H. Landfill Design for Reclamation of Şırnak Coal Mine Dumps—Shalefill Stability and Risk Assessment. In: ICMEMT; 11-12 July, 2014; Prag, Chekoslovakia; 2014

[20] Tosun Yİ. A case study on use of foam concrete landfill on landslide hazardous area in Şırnak City Province. In: XX Congress of the Carpathian Balkan Geological Association; Tirana, Albania; 24-26 September, 2014; 2014

[21] Tosun Yİ. Shale stone and Fly ash landfill use in land-slide hazardous area in Sirnak City with foam concrete. Geomaterials Journal. 2014;**4**(4): 141-150. DOI: 10.4236/gm.2014.44014

[22] Tosun Yİ. Benefaction from carbonation of flue gas CO_2 as coal mining filling. Geomaterials Journal. 2014;**4**(2):64-73. DOI: 10.4236/gm.2014.42007

[23] Tosun Yİ. Chemical activated grinding of Kutahya Gediz lignite, Şırnak asphaltite and shale by acidic coal mine water. In: 14th International Multidisciplinary Scientific GeoConference SGEM 2014, www.sgem.org, SGEM2014 Conference Proceedings; June 19-25, 2014. Book 1, Vol. 3; Albena Resort, Bulgaria; 2014. pp. 851-858. DOI: 10.5593/SGEM2014/B13/S4.108. ISBN: 978-619-7105-09-4/ISSN: 1314-2704

[24] Yıldırım İ. Tosun Kalker, Marn ve Şeylin Sünme Karakterizasyonu—Bitümlü Gözenekli Agrega için Don—Mikrodalga Kurutma-Bilya Darbe Dayanım Testi ile Sünme Etüdü. In: AGGRE 2016, 8th Internatıonal Aggregates Symposıum; October 5-7; Istanbul, Turkey; 2016

[25] Howard PH, Datta RS. Chemical cominution: A process for liberating the mineral matter from coal. In: Wheelock T, editor. Coal Desulfurization. ACS Series 64. Washington: American Chemical Society; 1976. pp. 58-64

[26] Howard PH, Datta RS, Hanchett A. Pre-combustion coal cleaning using chemical cominution. In: NCA/BCR Coal Conference and Expo Coal: Energy for Independence; Lousville; 1977

[27] Tosun Yİ. Use of modified freeze-drop ball test for investigation the crack propagation rate in coal mining—Case study for the şirnak asphaltite shale, marly shale and marl in şirnak coal site. In: IBSMTS 2016, 1th Internatıonal Black Sea Symposium on Mininig and Tunnelling; November 02-04; Trabzon, Turkey; 2016

[28] ASTM C 330. Standart Specifications For Lightweight Aggregates for Structural Concrete. Philadelphia: ASTM; 2013

[29] ASTM D6024-07. Standard Test Method for Ball Drop on Controlled Low Strength Materials. Pennsylvania: ASTM; 2007

[30] ASTM C 136. Standard Test Method for Sieve Analysis of Fine and Coarse Aggregates. Pennsylvania: ASTM; 2013

[31] Moulia M, Khelafib H. Performance characteristics of lightweight aggregate concrete containing natural pozzolan. Building and Environment. 2008;**43**:31-36

[32] Park CK, Noh MH, Park TH. Rheological properties of cementitious materials containing mineral admixtures. Cement and Concrete

Research. 2005;**35**:842-849. Available from: http://dx.doi.org/10.1016/j.cemconres.2004.11.002

[33] Piora LS, Piora IL. Production of expanded-clay aggregate for lightweight concrete from non-selfbloating clays. Cement and Concrete Composites. 2004;**26**:639-643

[34] Sarı D, Pasamehmetoglu AG. The effects of gradation and admixture on the pumice lightweight aggregate concrete. Cement and Concrete Research. 2005;**35**(5):936-942

[35] Şimşek O. Beton ve Beton Teknolojisi. I ed. Ankara: Seçkin Yayıncılık; 2004

[36] Szilagyi H, Terec L. Bricks recycled aggregates for structural green lightweight concrete. In: 13th SGEM GeoConference on Nano, Bio and Green Technologies for a Sustainable Future, SGEM2013 Conference Proceedings; 2013. pp. 375-380. DOI: 10.5593/sgem2013/bf6/s26.004. ISBN: 978-619-7105-06-3/ISSN: 1314-2704. Avialable from: www.sgem.org

[37] ASTM D616-07. Standard Test Method for Fly Ash Materials and Compositions, Slag Materials. Pennsylvania: ASTM; 2007

[38] TS EN 12390-3,-4. Beton-Sertlesmis Beton Deneyleri—Bölüm 3: Deney Numunelerinde Basınç Dayanımının Tayini. Ankara: Türk Standartları Enstitüsü; 2003

[39] Tosun YI. Chemical activated grinding of cement raw materials. In: Proceedings of XIVth International Mineral Processing Symposium; Kuşadası, Turkey; 2014

[40] Tosun YI. Co-polimer cementing of geothermal hot wells/discharge control in geothermal field, DGK 2017; Der Geothermie Kongress 2017; Eylül 12-14, 2017; Berlin, Germany; 2017

[41] Lakes RS, Kose S, Bahia HU. Analysis of high volume fraction particulates with applications to asphalt and high damping composites. Transactions of the ASME. 2002;**124**:174-178

[42] Masad E, Al-Rousan T, Bathina M, McGahan J, Spiegelman C. Analysis of aggregate shape characteristics and its relationship to hot mix asphalt performance. International Journal of Road Materials and Pavement Design. 2007;**8**(2):317-350

[43] Masad E, Little D, Sukhwani R. Sensitivity of HMA performance to aggregate shape measured using conventional and image analysis methods. International Journal of Road Materials and Pavement Design. 2004;**5**(4):477-498

[44] Masad E, Muhunthan B, Shashidhar N, Harman T. Aggregate orientation and segregation in asphalt concrete. ASCE Getotechnical Special Publication. 1998;**85**:69-80

[45] Masad E, Somadevan N, Bahia H, Kose S. Modeling and experimental measurements of strain distribution in asphalt mixes. Journal of Transportation Engineering. 2001;**127**(6):471-477

[46] Olard F, Perraton D. On the optimization of the aggregate packing characteristics for the design of high-performance asphalt concretes. Road Materials and Pavements Design. 2010;**11**(Suppl. 1):145-169. DOI: 10.3166/rmpd.11hs.145-169

[47] Tashman L, Masad E, Peterson B, Saleh H. Internal structure analysis of asphalt mixes to improve the simulation of superpave gyratory compaction to field conditions. Journal of the Association of Asphalt Paving Technologists. 2001;**70**:605-645

Chapter 7

Polymer-Modified Bio-Asphalt: A Sustainable Panacea to Greenhouse Gas Emissions

Modupe Abayomi Emmanuel

Abstract

The asphalt industry has been perpetually challenged with its health hazardous emissions and global warming issues, especially when produced with or from derivatives of fossil fuels and together with the need for higher strength to meet the heavy truck axle loads on highways. In view of the steady increase in high traffic intensity of heavily loaded trucks and the increase in the overloading of truck axles together with the significant variation in daily and seasonal temperature due to climate change effects, it is essential to modify the conventional asphalt cement using a combination of modifiers to improve the engineering properties of mixes to meet the complexity of the factors for higher strength and the environment. Global warming has consequently contributed immensely to adverse climate change effects, environmental degradation and attendant human health problems. The challenges posed by global pollution and depletion of fossil fuels leading to high cost of bituminous binder derivatives such as bitumen have motivated the search for and development and utilization of alternative binders, amongst which are polymer-modified bitumen and bio-oils. Research in this direction is hoping to find a permanent replacement for bitumen, thereby reducing greenhouse gas emissions and attendant climate change effects.

Keywords: bio-oil, bio-asphalt, sustainable pavements, pavement engineering, polymer-modified bitumen, polymer bio-asphalt, greenhouse gases

1. Introduction

The dynamics of world resource management suggest that all industries, including the asphalt pavement industry, should be exploring the benefits of economic, social and environmental sustainability in all spheres and sectors of development [19]. Sustainability is a growing concept of the twenty-first century Millennium Development Goals (MDGs), and it refers to the utilization of the Earth's resources in such a way that they fulfill the demands of the present generation without impeding the needs of future generations [1]. Sustainable pavements minimize the use of natural resources, reduce energy consumption, reduce greenhouse gas emissions, limit pollution, improve health and safety and ensure a high level of comfort for users [2]. For instance, the current high pavement construction costs have encouraged the use of wastes and renewable resources in asphalt pavements, with the eventual belief that a reduction in binder consumption shall have corresponding reduction in the cost of pavement mixes for development and maintenance [3]. The binder, bitumen, a major component in pavement mixes can be sourced from many avenues, amongst which are natural sources (lake asphalts) and by extraction from tar sands. It is a product of the fractional distillation of crude oil and is currently very expensive because of the high demand to meet the modern-day flexible pavement development for the rising number of automobiles on the roads, in both urban and rural areas [4].

Moreover, in recent years, the total price of asphalt mixes has increased significantly, owing to an increase in the price of the major binder and industrial fuel source, which is crude oil. Asphalt mixes are the most common and widely utilized materials in the construction and maintenance of transportation infrastructures. For instance, a 53% increment in the price of asphalt was reported by [5] to have occurred between the year 2004 and 2007, specifically from $68 to $104 per ton, over a 3-year span. Consequently, the attention of many highway pavement experts has been drawn to seek more economically and environmentally sustainable ways through the investigation of the means of cutting down on bitumen consumption in the asphalt mixes with alternative binders [6]. Besides, for sustainable pavement works, this approach can also produce sustainable binders, for example, from biomass sources, that can reduce the viscosity; reduce the demand for and consumption of virgin bitumen; lower the mixing and compaction temperature of the asphalt mixtures, as an attempt to address the environmental hazards associated with bitumen; and also reduce the heat emissions that mainly comes from the asphalt industry, which is believed to be worsening the effects of global warming [7]. The demand for bio-based fuels is gaining popularity as it is not only sustainable; it also provides secure energy and a positive economic opportunity [8]. The blend of bitumen with the alternative binders (bio-oils), obtained from vegetative or ornamental plants, is expected to improve the performance of bitumen and, by extension, polymer bio-asphalt with the inclusion of specialized strength improvement modifiers called polymers.

The focus of this paper is to discuss the global warming threat posed by the continuous dependence on fossil sources of pavement binders and evaluate the potentials of alternative binders especially polymer bio-asphalt for sustainable pavement works including the possibility of developing a polymer bio-based economy for developing countries as practiced currently in some developed nations of the world.

2. The asphalt industry and global warming threat

The asphalt industry is constantly challenged by its health hazardous emissions and global warming issues, especially when produced from dependence on fossil fuel sources for energy, industrial, commercial and domestic uses, together with the need for higher strength to meet the heavier truck axle loads on highways. Global warming has consequently contributed immensely to adverse climate change effects, environmental degradations and attendant human health problems in some developing economies around the world [9]. These challenges posed by global pollution and depletion of fossil fuels leading to high cost of bituminous binder derivatives such as bitumen have motivated the search for and development and utilization of alternative binders, amongst which are polymer-modified bitumen and bio-oils [10], which are blended together to produce polymer bio-asphalt. To reduce this global threat to climate and human health coupled with the need to improve the engineering properties of fossil-based asphalt pavements, selected polymer modifiers which include waste thermoplastic polyurethane, waste polyethylene terephthalate (PET) and waste tyre rubber and so on and bio-oils are blended with bitumen to produce a new hybrid polymer bio-asphaltic concrete with necessary investigations on modification effect on its mechanical and durability properties.

3. Background on polymer bio-asphalt constituents

3.1 Bitumen

Bitumen is a very complex, viscoelastic, rheological and non-crystalline material (black or brown in color), which is substantially soluble in carbon disulphide (CS_2) and exhibits adhesive and waterproofing characteristics. It is composed of aromatic hydrocarbons and particularly includes 80% of carbons (C) approximately, 15% of hydrogen (H) and remnants such as oxygen, sulfur, nitrogen, metals, etc. [11]. Bitumen however causes severe impact on our environment as well as the health of pavement workers when processed by heat. This is because of the release of vapor when it is heated to a temperature of about 165–200°C, which reduces the viscosity to a concentration that can coat all the aggregates for the desired construction purposes. The vapor released condenses after it has cooled down. It contains greenhouse gases and volatile components which are toxically and chemically different from the origin material. The fumes of bitumen contain tiny particles which are created as condensation occurs from its gaseous state [12]. These particles in the form of vapor mix with the air consumed by construction workers as well as other environmental elements resulting in toxic pollution. Bitumen is also an important low-cost thermoplastic which finds many applications as a building and engineering material; however, it is challenged with poor mechanical properties as it is hard and brittle in cold environments and soft and fluid in hot environments [13].

3.2 Bio oil

Bio-oil is a dark brown liquid with a smoke-like odor, produced from fast pyrolysis of biomass. Unlike the derivatives of petroleum, it has a different composition. The distribution of the constituent compounds varies, depending on the type of biomass and the intensity of the heat applied in its processing. Bio oil has a major component which is water. The other compounds are hydroxyl aldehydes, hydroxyl ketones, sugars, carboxylic acids and phenols [14, 15] used swine manure to produce oil. They successfully converted swine manure to oil with 50% efficiency at 380°C and 40 MPa (around 400 atm) pressure. Zhang et al. [16] reported that bio-oil possesses similar components as asphalt binder derived from petroleum. The components can be classified into asphaltenes, polars, aromatics and saturates. Nevertheless, as a result of the higher content of oxygen in bio oil, its elemental components also differ significantly from conventional asphalt [17]. Fini [18] in his research produced bio-oil from swine manure and used it as a partial replacement of bitumen. Thermochemical liquefaction process was used to convert swine manure to bio oil, while the remaining heavy residue obtained in the process was used as an asphalt modifier. In a research carried out by Raouf and Williams [19] through which bio-oil was produced from three sources, viz., corn stover, switch grass and oakwood, the bio-oil was mixed with bitumen and also blended with a polymer modifier. Results show a great improvement on the temperature susceptibility of the resultant mix after testing.

3.2.1 Biomass sources for bio-oil production

Biomass is any living matter on earth in which solar energy is stored. By the process of photosynthesis, plants produce biomass continuously [20]. Bio-fuels which are classified as liquid or gaseous fuels are produced from plant matter and residues of agricultural crops, municipal wastes and agricultural and forestry by products [21]. According to [22], biomass resources can be divided into two broad categories, that is, natural and derived materials, and then subdivided into three categories:

1. Wastes from agricultural produce, agricultural processing wastes, wastes from crop residues, wastes from mill wood, urban wood and organic wastes.

2. Wastes generated from forest products such as trees, shrubs, wood residues, bark and sawdust.

3. Energy crops and energy crop wastes such as grasses, sugar crops, oilseed crops, starch crops and herbaceous woody crops.

Biomass is the major energy source in Nigeria, contributing nearly 78% of Nigerian primary energy supply [23]. Woody and herbaceous species, wood wastes, energy crops, bagasse, agriculture and industrial residues, waste paper, municipal solid waste, sawdust, bio solids, grass, waste from food processing, animal wastes, aquatic plants and algae, etc. are different kinds of biomass resources [24]. Specifically, in Nigeria, the following biomass resources are available: fuel wood, agricultural waste and crop residue, sawdust and wood shaving, animal dung/poultry droppings, industrial effluents and municipal solid wastes [26]. Others which have been recently discovered and reported to possess high potentials to be used as binders are *Jatropha curcas*, *cassava peels* and *sawdust*. **Table 1** shows the estimated biomass resources in Nigeria. Biomass resources are essentially composed of organic polymers such as cellulose, hemicellulose and lignin, and they include various natural and derived materials.

3.2.1.1 Jatropha curcas

Jatropha curcas (physic nut) belonging to the *Euphorbiaceous* family, a native of tropical America, has been introduced into Africa and Asia and is now cultivated worldwide over an area of approximately 1 million ha [27]. Studies and forecasts by the Global Social Investment Exchange indicated a strong expansion in the cultivation of this crop by up to 12.8 million ha in 2015 [28]. Another study by the International *Jatropha* Organization confirmed the trend, with a forecast of 160 million tons of seeds from 32.72 million ha cultivated worldwide in 2017. *Jatropha* is a genus of approximately 175–200 plants, shrubs and trees adapted to arid conditions, which can easily be propagated by cutting, and is extensively planted with a built-in capacity to combat desertification by restoring vegetative cover. Its seeds contain some proportion of crude oil, proteins, water, crude fibers, ash and carbohydrates. The plant is also relatively drought resistant and has potential for controlling soil erosion and increasing the habitat of wild animals. The plant does not require any particular soil type for growth, can flourish on almost any soil composition and can produce seeds containing up to 40% mass of oil [29]. Jatropha has become a very important crop to the developing world that is constantly challenged by the effects of climate change due to its ability to

Resources	Quantity (million tons)	Energy value ('000 MJ)
Fuel wood	39.1	531
Agro-waste	11.244	147.7
Sawdust	1.8	31.433
Municipal solid waste	4.075	—

Source: Kuye and Ede, 2013; [25].

Table 1.
Biomass resources and the estimated quantities in Nigeria in the year 2000.

grow on arid, waste or marginal lands, producing energy crops without displacing food crops. Its ability to reclaim problematic lands and restore eroded areas is added advantages. *Jatropha* plays an important role in keeping out cattle and protecting other valuable food crops or cash crops. *Jatropha* products from the fruit—the flesh, seed coat and seed—cake are rich in nitrogen, phosphorous and potassium and are fertilizers that improve soil. Biodiesel extracted from *Jatropha* can be used for rural electrification, transportation and in industries [30]. Thus, *Jatropha curcas* oil is considered a more sustainable feedstock for energy production than any other food-based crop, such as palm oil, rapeseed oil, soybean oil and sunflower oil [31].

Besides, the increasing emphasis on pavement sustainability and the promotion of the use of alternative resources as binder, coupled with the successes achieved in using waste engine oil (WEO) and waste cooking oil (WCO) in rejuvenating recycled asphalt pavements, has driven the idea of using *Jatropha curcas* oil (JCO), a nonedible oil as a result of the presence of anti-nutritional factors, such as phorbol esters as a bio-based rejuvenating agent for aged bitumen. This means that it possesses resistance to aging and has good storage stability, making it a focus of interest in this research in addition to its availability and low cost. Several vegetable oils have been used as raw materials for biodiesel production and in flexible pavement applications. Vegetable oils such as palm oil, soybean oil, sunflower oil, coconut oil, rapeseed oil, tung oil, groundnut and palm kernel oils which are classified as edible oil seeds have been used. Even the use of oils from algae, microalgae, bacteria and fungi has been investigated [32]. Some of these feedstocks are found in abundance in Nigeria. The nonedible ones that are most commonly found are *Jatropha curcas* (*Jatropha curcas* L.) and neem (*Azadirachta indica*), and very little or no research especially in flexible pavement applications has been carried out [29] investigated the feasibility of utilizing JCO as a bio-based recycling and rejuvenating agent through physical tests and storage stability properties of bitumen. The results showed that JCO was capable of rejuvenating and recycling aged bituminous mixtures as well as reducing construction and maintenance cost. This conclusion was derived from the positive test results on penetration index (PI) which is an indication of good resistance to low temperature cracking and storage stability and increase in softening point.

Consequent upon these exciting and inherent benefits of the utilization of *Jatropha* oil and because there has been no research until now that studies its modification of fresh bitumen, it is the interest of this study to investigate the applicability of utilizing it as partial replacement for bitumen in production of asphaltic concrete for sustainable pavement works.

3.2.1.2 Cassava feedstock

Cassava, the third largest carbohydrate source in the world, is widely cultivated as an annual crop in tropical and subtropical regions. This is due to its starchy and edible tuberous root. It has heavy presence in countries like Thailand, Brazil and Nigeria and is a very promising feedstock for ethanol production [33]. Residues from cassava plantations include rhizome, stalk and peels. Nigeria is greatly endowed with fertile hectares of land that support the cultivation of agricultural biomass. Agricultural biomass is a relatively broad category of biomass that includes the food-based portion of crops and nonfood-based portion of crops, as seen in the energy crop presented above. The food-based portion of crops is the part of the plant that is either oil or simple sugars. Examples include rapeseed (used for canola oil), sunflower, soybeans, corn, sugarcane and sugar beets. Corn and sugar beets can be used for ethanol production through fermentation, while the oilseed crops can be used to produce biodiesel. On the other hand, the nonfood-based portion of crops is that part which is usually discarded as a residue. Crop residues are those residues that are associated with agriculture either as on-the-farm crop wastes, such as cornstalks or as processing wastes such as rice husk, corn shells, palm kernel shell, cassava peels, cassava rhizome, cassava stalk, etc. [34]. They are available in Nigeria in large quantities especially in the rural areas where crops are cultivated. From **Table 1**, the agro-waste (crop residues) in Nigeria was estimated in the year 2000 to be about 11.244 million tons of the entire biomass resources of that year, and this amounted to 147,700 MJ of energy [35]. These wastes are currently burned directly as a starter or supplement materials to fuelwood. They have a great potential for bio-fuel production. Recently, some of these biomasses have been used by different researchers to produce bio oil. These include rice husks and sawdust [30]and cassava plantation like cassava stalk and cassava rhizome [36] and cassava peels [37].

[37] reported in their experiment that bio-oil yield from fast pyrolysis of cassava peel ranged from about 38.7 to 51.2%. The maximum yield of bio-oil was obtained at temperature of 525°C. The main organic composition of bio-oil can be grouped into acids, esters, aldehydes, ketones, phenols, alcohols, hydrocarbons and other organic compounds and some un-identified organic compounds. The physicochemical properties of bio-oil produced from pyrolysis of cassava peels were comparable to the ASTM standard for bio-oil. However there has been no research until now that studies the applicability of utilizing bio oil from cassava feedstock as partial replacement for bitumen for sustainable pavement works.

3.2.1.3 Sawdust

Sawdust is a typical example of lignocellulosic biomass (biomass that is mostly plant cell walls which have high carbon content). It is readily available in Nigeria, as can be seen from **Table 1**. It is associated with the lumber industry and the main by-product of sawmill. The sawdust generated from sawmills are often discarded, burnt as wastes or used for cooking but may also constitute an environmental hazard if not properly used. However, as noted earlier, sawdust can also be converted to bio-oil. In the recent times, [38] investigated the use of sawdust in the production of bio oil. Also, [39] investigated fast pyrolysis of waste furniture sawdust in a fluidized bed. Specifically, large quantities of oxygenated compound are present in bio-oils produced from wood. In the bio-oils huge proportion of xylenols, eugenols, cresols and phenols are also found. It has been observed that the phenolic fraction has shown good performance as an adhesive for waterproof plywood as reported by [20]. Nigeria has a great potential for the various kinds of biomass, such as sawdust, agricultural residue, wood crop and municipal waste. Sawdust has been recognized as a biomass with the highest yield of bio oil according to [40]. This has prompted research into the sustainability potential of utilizing it as bio modifier for bitumen for use in flexible pavement, especially sustainable pavement construction.

3.2.2 Production of bio-oils by pyrolysis

Pyrolysis is a technology used to extract bio-oil from biomass. It produces three different products: syngas, biochar and bio oil. It is a process that has been found to be versatile, efficient and environmentally acceptable [30]. The conversion process from biomass to bio-oil is through fast pyrolysis, slow pyrolysis, carbonization, liquefaction or gasification. Fast pyrolysis is the technique most widely used because of its many advantages, such as high yield, simplicity and low cost [41]. In the absence of oxygen, biomass decomposes at a temperature of 400–600°C, and during this process, a high energy product known as bio-oil is produced which has the capacity to replace conventional fuels and can be applied in different ways, such as in boilers and turbines and high-power diesel motors. According to [20] three products are produced during pyrolysis, which is conducted in the absence of oxygen; they are a mixture of noncondensable gases (syngas), a charcoal coproduct and a condensable liquid (pyrolysis oil). He further revealed that through pyrolysis, a complex mixture of oxygenated compounds is formed by the conversion of the organic materials.

3.2.3 Physical and rheological properties of bio-oils

The capability of bio-oil utilization as bio-binders, including its reliability, has been discovered to be properly investigated through physical property measurements as chemical analysis has proven to be very difficult in terms of performance characterization. According to a report by [42], the following is the description of the physical state of bio-oils: "Bio-oils' multiphase complex structure is accounted for by the presence of aqueous droplets with different natures, waxy materials, char particles, and micelles formed from heavy molecular compounds in a matrix of hollo cellulose-derived compounds and water". During transportation, and sometimes when stored or processed, bio-oil constituent compounds such as ketones, aldehydes and others can react through condensations to form larger molecules [21]. Therefore, these reactions lead to the undesirable changes in physical properties. For example, viscosity and water content can increase, whereas the volatility will decrease [43]. As reported by [44], this is analogous to the behaviour of asphaltenes contained in petroleum by some means. The following are the summarized physical properties of bio-oils: (1) a density of 1200 kg/m^3 higher than the original biomass; (2) a viscosity that ranges from 25 cPoise to 1000 vis-à-vis moisture content, the amount of light compounds and the aging [20]; and (3) a water content which ranges between 14 and 33% by weight. Williams et al. [45, 46]opined that the rheological properties play a significant role in describing the behaviour of bio-oils. Measuring the rheological properties is useful to determine behavioral and predictive information for bio-oils as well as knowledge of the effect of processing, formulation changes and aging phenomena.

3.2.4 Chemical properties of bio-oils

Bio-oils generally possess a complex chemical nature, thus making it difficult to conduct a complete chemical characterization, and sometimes it is almost impossible to do so. According to the findings by [43], the complexity of chemical analysis usually results from the presence of high molecular weight of phenolic species from the decomposition of lignin. Besides, different numbers of carboxylic and phenolic acids, hydroxyl groups and aldehydes exist alongside the fragmented oligomeric products, including alcohol and ether functions. Thus, phenolic species exist as different hydrogen-bonded aggregates, micelles, droplets and gels. Due to the complexity of the chemical structure and the broadness of chemical properties of bio-oils, the few chemical indices that are commonly investigated are corrosiveness, distillation, homogeneity, water content, molecular weight, oxidation and aging and phase stability.

3.3 Polymer modifiers

3.3.1 Definition of polymer

Polymer is a combination of two classical Greek words poly and meres which literally means "many" and "parts" [47]. It is a molecule with a long chain, consisting of many repeating units that are identical in structure. Polymers can be classified in so many ways, according to their origin (whether they are natural or synthetic (man-made)). The natural polymers such as cellulose, hemicellulose and lignin are found in nature; while the synthetic polymers are man-made polymers that are formed through synthetic routes, such as waste plastics (polyethylene) and crumb rubber (CR) (vulcanized rubber or scrap tyres), to mention a few. Plastics are low-cost, low-density (or light-weight), formable and durable materials which, due to their properties, are extensively used in many areas, sectors and industries. Plastics are of seven types: PET, high-density polyethylene (HDPE), polyvinyl chloride (PVC), low-density polyethylene (LDPE), polypropylene (PP), polystyrene (PS) and other types of plastics [48].

3.3.2 Significance of polymer modifiers

To obtain a bio-binder with suitable and similar binding characteristics for use as substitute for conventional bitumen, there is a need for the upgrading of the bio-oil produced from fast pyrolysis of biomass. According to previous studies, the bio-oil by itself possesses very good characteristics, like asphalt, at high and medium temperatures [19]. However, the performance at low temperatures was not acceptable for use in most of the USA, as they were too stiff at low temperatures and susceptible to low-temperature cracking. The approach used to solve this issue was to react a polymer with the bio-oil. A bitumen additive should improve binder properties at both low and high in-service temperatures [49]. Consequently, at high temperatures, it should be durable enough to resist loads from traffic, which may result into rutting or permanent deformation; and also at low pavement temperatures, it is expected to be flexible enough to avoid excessive thermal stresses. It also needs to be able to endure the thermal cycling of service without cracking or deforming. Furthermore, a modifying agent of bitumen should be easily incorporated to yield a highly viscous mixture at in-service temperatures which remains homogenous on storage and should have a viscosity which permits its use in standard material manufacturing and paving equipment. Besides, it should be highly resistant to ultraviolet light, thermal action and water, not leach deleterious substances into the environment, and be readily available [42].

3.3.3 Modification of bituminous mixes by crumb rubber and plastic wastes (PET)

Polymer modifiers have been reported to possess the capacity to improve the properties of bitumen and thus the overall properties of hot-mix asphalt (HMA) used in flexible pavement works [50]. Since the early 1970s, the utilization of petroleum-derived polymers has been well developed to be blended with conventional bituminous binders to modify the performance and rheological properties by decreasing temperature susceptibility and increasing cohesion as reported by [51]. In other words, practical experience has shown that the blending of bitumen binders with polymer modifiers (e.g. polyethylene and crumb rubber) has many advantages that include but are not limited to enhanced fatigue resistance, improved thermal stress cracking, decrease in temperature susceptibility and reduction of rutting [52]. Polymers that have been commonly used to modify bitumen include styrene-butadiene-styrene (SBS) copolymer, styrene-butadiene rubber (SBR), ethylene-vinyl acetate (EVA), polyethylene terephthalate (i.e. LDPE, HDPE, etc.) and waste polymers (plastic from agriculture, crumb tyre rubber, etc.) [53]. For such polymers, the mixing process may have a significant effect on the technical properties of the resulting blend, as well as on the costs of the whole operation [48]. Thus, to bring down the differences in viscosity between bitumen and polymer and to obtain suitable dispersions of polymer, high shear and processing temperatures (170–180°C) are necessary. Therefore, the final binder may undergo a "primary" aging, mainly due to the oxidation of maltene compounds and polymer degradation, leading to a decrease in the expected mechanical performance of bituminous binders [45]. Moreover, the use of high-molecular-weight polymers may yield thermodynamically unstable modified bitumen, and phase separation readily occurs during their storage at higher temperatures [42]. Besides, using PET-derived products as a performance-enhancing additive for asphalt can be one of the alternatives to recycle and reuse this waste material. Similarly, disposal of waste vehicle tyres is also a challenging environmental problem. Research has shown that the use of crumb rubber from waste tyres in asphalt helps improve some of its rheological properties but is often plagued with other concerns such as low storage stability. Leng et al. [28] in their research found out that the incorporation of PET-based additives to CRMA improved the storage stability, rutting and fatigue resistances and increased the rotational viscosity (RV) of the modified binders.

Generally, tyre rubber comprises of natural, synthetic rubbers, carbon black and other mineral fillers. Crumb rubber-modified asphalt can be divided into two categories, the wet and dry processes. The wet process is the most efficient in improving properties of an asphalt mixture [54]. Crumb rubber is known to absorb liquids and swell, depending on the temperature and viscosity of the liquids it is absorbing. Crumb rubber is a general term used to describe a granular rubber from waste tyres. At high temperatures, the asphalt binder tends to flow easier due to the natural decrease of viscosity associated with high temperatures. This condition creates a softer asphalt mixture, which is prone to rutting. The addition of crumb rubber to the HMA provides extra viscosity, thus stiffening the HMA at the high temperatures [54].

4. Conclusion

Modification of base bitumen with polymer and bio-oil blends or polymer bio-asphalt, to enhance the main engineering properties of base bitumen such as viscosity, penetration, softening point, flashpoint, fire point, density, specific gravity and so on, including Marshall properties of resultant polymer bio-asphalt concrete, can be a sustainable way out of the global warming crises and climate change experienced all over the world; therefore, it should be tenaciously promoted given its outstanding developmental and economic benefits. Furthermore, it is recommended that optimum combination proportions of both bio-oil from cassava peel and crumb rubber be investigated. The use of bitumen modified with bio-oil and crumb rubber for use in the construction of flexible pavements is recommended to reduce agricultural wastes and other environmental pollution issues as well as reduce road construction costs.

Acknowledgements

The author appreciates the Almighty God and the support of colleagues.

Conflicts of interest

Author declares no conflict of interest.

Funding

This work has not received funding from any source whatsoever.

Author details

Modupe Abayomi Emmanuel
Department of Civil Engineering, Highway and Transportation Engineering Research Group, Landmark University, Nigeria

*Address all correspondence to: modupe.abayomi@lmu.edu.ng

References

[1] Aziz MA, Rahman T, Rosli M, Azelee W, Abu W. An overview on alternative binders for flexible pavement. Construction and Building Materials. 2015;**84**:315-319

[2] Indexed S, Olumoyewa D, Aran O. Development and performance evaluation of crumb rubber—Bio-oil modified hot mix asphalt for sustainable highway pavements. International Journal of Civil Engineering and Technology. 2019;**10**(2):273-287

[3] Modupe AE, Atoyebi OD, Oluwatuyi OE, Aladegboye OJ, Busari AA, Basorun AO. Dataset of mechanical, marshall and rheological properties of crumb rubber—Bio-oil modified hot mix asphalt for sustainable pavement works. Data Brief. 2018;**21**:63-70

[4] Williams RC, Cascione AA, Cochran EW. Development of Bio-Based Polymers for Use in Asphalt; 2014

[5] Jamshidi A, Othman M, You Z. Performance of warm mix asphalt containing sasobit Ò: State-of-the-art. Construction and Building Materials. 2013;**38**:530-553

[6] Poh CC, Hassan NA, Azah N, Raman A. Effect of Fast Pyrolysis Bio-Oil from Palm Oil Empty Fruit Bunch on Bitumen Properties; 2018

[7] Williams RC, Rover M, Brown RC. Utilization of Fractionated Bio-Oil in Asphalt; 2011

[8] Yang X, You Z. High temperature performance evaluation of bio-oil modified asphalt binders using the DSR and MSCR tests. Construction and Building Materials. 2015;**76**:380-387

[9] Owamah HI, Izinyon OC. Bioresource technology development of simple-to-apply biogas kinetic models for the co-digestion of food waste and maize husk. Bioresource Technology. 2015;**194**:83-90

[10] Abdel M, Mohamed R. Development of Non-petroleum Binders Derived from Fast Pyrolysis Bio-Oils for Use in Flexible Pavement; 2010

[11] Raman NAA et al. A review on the application of bio-oil as an additive for asphalt. Jurnal Teknologi. 2015;**72**(5):2017

[12] Poh CC, Hassan NA. The Effect of Oil Palm Waste Pyrolysis Bio-Oil on Modified Bitumen Properties

[13] McNally T editor. Polymer modified bitumen: Properties and characterization. Elsevier; 2011

[14] Gobinath R, Ganapathy GP, Akinwumi II, Kovendiran S, Hema S, Thangaraj M. Plasticity, strength, permeability and compressibility characteristics of black cotton soil stabilized with precipitated silica. Journal of Central South University. 2016;23(10):2688-2694

[15] Jensen EA, Mousavi M, Pahlavan F. Multiscale investigation of a bioresidue as a novel intercalant for sodium montmorillonite. The Journal of Physical Chemistry C. 2017;**121**(3):1794-1802

[16] Zhang R et al. Thermal storage stability of bio-oil modified asphalt. Journal of Materials in Civil Engineering. 2018;**30**(4):1-9

[17] Zhang R, Wang H, Gao J, You Z, Yang X. High temperature performance of SBS modified bio-asphalt. Construction and Building Materials. 2017;**144**:99-105

[18] Fini EH et al. Source dependency of rheological and surface characteristics of bio-modified asphalts. Road

Materials and Pavement Design. 2016;**18**:408-424

[19] Raouf MA, Williams RC. Temperature and shear susceptibility of a nonpetroleum binder as a pavement material. Transportation Research Record. 2010;**2180**(1):9-18

[20] Demirbas MF. Recent advances on the production and utilization trends of bio-fuels: A global perspective. Energy Conversion and Management. 2006;**47**:2371-2381

[21] Mohan D, Pittman CU, Steele PH. Pyrolysis of wood/biomass for bio-oil: A critical review. Energy & Fuels. 2006;**20**(3):848-889

[22] Goyal HB, Seal D, Saxena RCÃ. Bio-fuels from thermochemical conversion of renewable resources: A review. Renewable and Sustainable Energy Reviews. 2008;**12**:504-517

[23] Agbro EB, Ogie NA, Rjeas R. A comprehensive review of biomass resources and biofuel production potential in Nigeria. Research Journal in Engineering and Applied Sciences. 2012;**1**(3):149-155

[24] Yaman S. Pyrolysis of biomass to produce fuels and chemical feedstocks. Energy Conversion and Management. 2004;**45**:651-671

[25] Kuye AO, Edeh I. Production of bio-oil from biomass using fast pyrolysis: A critical review. Journal of Minerals Research. 2013;**1**(1):1-19

[26] Sambo BAS. Strategic developments in renewable energy in Nigeria. International Association for Energy Economics. 2009;**16**(3):15-19

[27] Simón D, Borreguero AM, De Lucas A, Gutiérrez C, Rodríguez JF. Sustainable polyurethanes: Chemical recycling to get it. In: Environment, Energy and Climate Change I. Springer, Cham; 2014;229-260

[28] Leng Z, Padhan RK, Sreeram A. Production of a sustainable paving material through chemical recycling of waste PET into crumb rubber modified asphalt. Journal of Cleaner Production. 2018;**180**:682-688

[29] Ahmad KA, Abdullah ME, Hassan NA. Investigating the Feasibility of Using *Jatropha curcas* Oil (JCO) as Bio Based Rejuvenator in Reclaimed Asphalt Pavement (RAP)2017. p. 09013

[30] Zhu J et al. GC – MS and FT-IR analysis of the bio-oil with addition of ethyl acetate during storage. Construction and Building Materials. 2014;**2**(2):53-60

[31] Subroto E, Manurung R, Heeres HJ, Broekhuis AA. Mechanical extraction of oil from *Jatropha curcas* L. kernel: Effect of processing parameters. Industrial Crops and Products. 2015;**63**:303-310

[32] Aransiola EF, Daramola MO, Ojumu TV, Aremu MO, Layokun S, Solomon BO. Nigerian *Jatropha curcas* Oil Seeds: Prospect for Biodiesel Production in Nigeria. Vol. 2; 2012

[33] Nigeria to Produce Bioethanol from Cassava; Biofuels International Magazine

[34] Edhirej A, Sapuan SM, Jawaid M, Zahari NI. Cassava: Its polymer, fiber, composite, and application. Polymer Composites. 2017;**38**(3):555-570

[35] Raman NAA, Hainin MR, Hassan NA, Ani FN. A review on the application of bio-oil as an additive for asphalt. Jurnal Teknologi. 2015;**72**(5)

[36] Pattiya A, Titiloye JO, Bridgwater AV. Catalytic Pyrolysis of Cassava Rhizome. Table 2 Characteristics of Selected Catalyst Samples. Vol. 202006. pp. 1-6

[37] Ismadji S, Ju Y, Xiang C, Kurniawan A, Lu O. Bio-Oil from Cassava Peel:

Potential Renewable Energy Source; 2012

[38] Chang S et al. Bioresource technology effect of hydrothermal pretreatment on properties of bio-oil produced from fast pyrolysis of eucalyptus wood in a fluidized bed reactor. Bioresource Technology. 2013;**138**:321-328

[39] Su H et al. Bioresource technology bio-oil production from fast pyrolysis of waste furniture sawdust in a fluidized bed. Bioresource Technology. 2010;**101**(1):S91-S96

[40] Wei L, Cheng S, Zhao X, Huang Y, Yu Y. Exploration of Lignocellulosic Biomass Precision Pyrolysis for Advanced Biofuel Production. In 2015 ASABE Annual International Meeting American Society of Agricultural and Biological Engineers. 2015. p. 1

[41] Carpenter D, Westover TL, Czernik, S, Jablonski W. Biomass feedstocks for renewable fuel production: A review of the impacts of feedstock and pretreatment on the yield and product distribution of fast pyrolysis bio-oils and vapors. Green Chemistry. 2014;**16**(2):384-406

[42] Garcia-Perez M et al. Fast pyrolysis of oil mallee woody biomass: Effect of temperature on the yield and quality of pyrolysis products. Industrial and Engineering Chemistry Research. 2008;**47**(6):1846-1854

[43] Nanda S, Mohanty P, Kozinski JA, Dalai AK. Physico-chemical properties of bio-oils from pyrolysis of lignocellulosic biomass with high and slow heating rate. Energy and Environment Research. 2014;**4**(3):21-32

[44] Czernik S, Bridgwater AV. Overview of applications of biomass fast pyrolysis oil. Energy & Fuels. 2004;**18**(2):590-598

[45] Raouf MA, Engineering T, Williams RC. Determination of Pre-Treatment Procedure Required for Developing Bio-Binders from Bio-Oils; 2009

[46] Raouf MA, Williams CR. General rheological properties of fractionated switchgrass bio-oil as a pavement material. Road Materials and Pavement Design. 2011;**11**(sup1):325-353

[47] Fried JR, Hu N. The molecular basis of CO_2 interaction with polymers containing fluorinated groups: Computational chemistry of model compounds and molecular simulation of poly [bis(2,2,2-trifluoroethoxy) phosphazene]. Polymer. 2003;**44**: 4363-4372

[48] Habib NZ, Kamaruddin I, Napiah M, Tan IM. Rheological properties of polyethylene and polypropylene modified bitumen. International Journal of Civil and Environmental Engineering. 2010;**4**(12):381-385

[49] Polacco G, Vacin OJ, Biondi D, Stastna J, Zanzotto L. Dynamic master curves of polymer modified asphalt from three different geometries. Applied Rheology. 2003;**13**(3):118-124

[50] Köfteci S, Ahmedzade P, Kultayev B. Performance evaluation of bitumen modified by various types of waste plastics. Construction and Building Materials. 2014;**73**:592-602

[51] Aisien FA, Hymore FK, Ebewele RO. Application of ground scrap Tyre rubbers in asphalt concrete pavements. International Journal of Materials Science and Engineering. 2006;**13**:333-338

[52] Totry E, González C, Llorca J. Mechanisms of shear deformation in fiber-reinforced polymers: Experiments and simulations. International Journal of Fracture. 2009;**158**(2):197-209

[53] El-ghaffar MAA, Youssef EAM. Maleic acid/phenylene diamine adducts as new antioxidant amide

polymers for rubber (NR and SBR) vulcanizates. Polymer Degradation and Stability. 2003;**82**:47-57

[54] Panda M, Mazumdar M. Utilization of reclaimed polyethylene in bituminous paving mixes. Journal of Materials in Civil Engineering. 2002;**14**(6):527-530